Carbon-Neutral Pathways for China: Economic Issues

Kai Tang
Editor

Carbon-Neutral Pathways for China: Economic Issues

 Springer

Editor
Kai Tang ⓘ
Guangdong University of Foreign Studies
Guangzhou, China

ISBN 978-981-19-5564-8   ISBN 978-981-19-5562-4 (eBook)
https://doi.org/10.1007/978-981-19-5562-4

This Springer imprint is published by the registered company Springer Nature Singapore Pte Ltd.
The registered company address is: 152 Beach Road, #21-01/04 Gateway East, Singapore 189721,
Singapore

Introduction

The global communities nowadays widely support that it is vital to achieve carbon neutrality as early as possible. Growing scientific evidence argues that uncontrolled anthropogenic emissions of greenhouse gases (GHG), mainly carbon dioxide (CO_2) emissions associated with fossil fuel use, are likely to bring increasingly frequent fires, droughts, flooding, and 1.5 °C warming about 2030. Therefore, deep and fast cuts in GHG emissions, which could stabilize rising temperatures and avoid the possible catastrophe, are in great demand. Accordingly, a growing number of countries have announced their carbon neutrality plans. In 2020, some major economies announced target dates for achieving carbon neutrality, many aiming for 2050 (e.g., Japan, Germany, and Canada).

China is the largest emitter of GHG in the world. Every year, China emits more GHG than the entire developed countries combined. In 2019, China emitted 27% of the world's GHG, while the US and India contributed 11% and 6%, respectively. Under both international and domestic pressures to do more to address global warming, China's leaders have recently pledged to peak its emissions within the next 10 years and become carbon neutral before 2060. This is the first long-term climate goal of the world's most populous country, which could affect more than the country's 1.4 billion people. Considering China's ongoing rapid industrialization and urbanization, the goal is ambitious and challenging.

Despite China's efforts to push for carbon neutrality, its pathways are yet not clear, especially from an economic perspective. Scientific studies suggest that a balanced mixture of strategies aiming to reduce and offset GHG emissions is needed to achieve carbon neutrality. Establishing carbon neutrality pathways needs to identify the best combination of these strategies by considering their costs and benefits, tradeoffs, interdependence, and impacts to economic development, social welfare, and environmental quality. Furthermore, carbon neutrality pathways need to be customized for different actors ranging from producers, industries, regions, and the country. Carbon neutrality goals and pathways of these actors might contradict each other.

The economic issues on China's carbon-neutral pathways have rarely been discussed and documented either on media or in the country's policy. There are also a very limited number of books available on carbon-neutral pathways assessment in China. In fact, there is not a single comprehensive reference book available on the economic issues of this subject. Most of the research is scattered either in the form of publications or in the form of reports on related scientific and engineering issues.

What This Book is About

This book is the first comprehensive assessment of carbon-neutral pathways in China from an economic perspective. This book gives a detailed overview of issues and challenges related to carbon neutrality which have not been adequately addressed, e.g., reduction costs and efficiency of existing actions, the multiple impacts of the newly established carbon market, the potentials and costs of nature-based solutions such as biophysical sequestration, etc. Studies on China's carbon reduction have attracted scientists and policymakers from diverse backgrounds. Pursuing a holistic and systematic approach, the book establishes a fundamental framework for this topic, emphasizing the importance of integrated technical-economic-policy analysis.

Who This Book is For

Given the literature dearth, this book is not only a comprehensive assessment of carbon neutrality in China but also an outstanding text book on carbon-neutral management in China. Similarly, this book is expected to provide a reference for a great range of readership including undergraduate and postgraduate students, economic and climate specialists, researchers and policymakers in China as well as in overseas.

How This Book is Organized

The following is an outline of this book. If you are not familiar with the related areas, I recommend you read the chapters in order. Nevertheless, each chapter is relatively self-contained and can be read individually if you are interested in a particular topic.

Chapter 1 addresses why China's net-zero emissions matter to us. This provides a backcloth for the following chapters.

Chapter 2 explores spatiotemporal dynamics of China's carbon emissions by conducting a city-level convergence analysis. Chapter 3 analyzes the spatiotemporal dynamics of China's carbon emissions from industrial and regional decoupling perspectives. Those two chapters provide a picture of the country's carbon emissions.

Chapter 4 addresses provincial reduction potentials and reduction costs of carbon emissions. Chapter 5 focuses more closely on city-level carbon reduction costs and potentials. It is necessary to have a sound understanding of the costs and potentials associated with the carbon reduction efforts, which will contribute to designing and refining economic and climate policy regimes.

Chapter 6 conducts a whole-farm bioeconomic analysis to investigate the variations in enterprises, on-farm GHG emissions, and marginal abatement costs with various levels of market-based incentives. Agriculture is expected to make a great contribution to China's carbon-neutral progress.

Chapter 7 empirically investigates the impact of China's pilot carbon emission trading on urban carbon emissions. Chapter 8 explores the impact of China's pilot emissions trading on provincial industrial carbon emissions. Chapter 9 adopts a quasi-natural experiment to empirically investigate the impact of China's pilot ETS on industrial carbon productivity. It is useful to have a deeper understanding of the role and effectiveness of the global largest carbon market.

Chapter 10 conducts a comprehensive review of the literature on the strategies for agricultural carbon sequestration and GHG emission reductions and their economic feasibility. A general lesson from this review is that carbon sequestration and GHG emission reductions in agriculture are potentially attractive.

Prof. Dr. Kai Tang

Contents

1 China's Net-Zero Emissions: Why Do We Need It 1
Kai Tang

**2 Spatiotemporal Dynamics of China's Carbon Emissions:
Evidence from Urban Efficiency and Convergence** 19
Kai Tang and Di Zhou

**3 Spatiotemporal Dynamics of China's Carbon Emissions:
Evidence from Industrial and Regional Decoupling** 33
Lin Yang and Kai Tang

**4 Provincial Carbon Reduction Costs and Potentials in China:
A Total Factor Analysis** 49
Kai Tang and Lin Yang

**5 Urban Carbon Reduction Costs and Potentials in China:
A Nonparametric Approach** 65
Jianxin Wu and Kai Tang

6 Cost-Effectiveness of Agricultural Carbon Reduction in China 81
Kai Tang and Dong Wang

**7 Investigating the Impact of Carbon Emission Trading
on Urban Carbon Emissions in China** 95
Kai Tang and Yichun Liu

**8 Investigating the Impact of Carbon Emission Trading
on Provincial Industrial Carbon Emissions in China** 111
Kai Tang and Ye Zhou

**9   Investigating the Impact of Carbon Emission Trading
      on Industrial Carbon Productivity in China** 131
      Di Zhou and Kai Tang

**10   Carbon Sequestration and Greenhouse Gas Emissions
      Reductions in Agriculture: Strategies and Their Economic
      Feasibility** 149
      Kai Tang

About the Editor

Dr. Kai Tang received a Ph.D. degree in Agricultural and Resource Economics from the University of Western Australia in September 2016 and a Ph.D. degree in Natural Resources Management from China University of Geosciences in December 2016. He is now a Professor in School of Economics and Trade, Guangdong University of Foreign Studies. His expertise lies in interdisciplinary research, agricultural economics, environmental economics, and environmental modeling. Since 2016, he has published 30 peer-reviewed papers in international journals, including Agricultural Systems, Australian Journal of Agricultural and Resources Economics, Journal of Environmental Management, Land Use Policy, Energy, Applied Energy, Sustainable Development, etc., and two books. He is the Deputy Secretary-General of Green Finance Division, China Society of Natural Resources and a Member of Standing Committee of Energy & Resource Systems Engineering Division, Systems Engineering Society of China. He also serves as a reviewer of more than 30 international journals, i.e., Nature Energy, Ecological Economics, Climate Policy, Agricultural Systems, Journal of Environmental Management, Energy, etc.

Chapter 1
China's Net-Zero Emissions: Why Do We Need It

Kai Tang

1.1 Climate Change: Humanity's Biggest Challenge

Climate change refers to change in the long-term pattern of behaviour of the atmosphere caused by natural processes or anthropogenic activities. The Intergovernmental Panel on Climate Change (IPCC) defines climate change as "*a change in the state of the climate that can be identified (e.g., by using statistical tests) by changes in the mean and/or the variability of its properties and that persists for an extended period, typically decades or longer*" (IPCC 2018). The United Nations Framework Convention of Climate Change (UNFCCC) defines climate change as "*a change of climate that is attributed directly or indirectly to human activity that alters the composition of the global atmosphere and that is in addition to natural climate variability observed over comparable time periods*" (UNFCCC 2011). The UNFCCC definition therefore notices the distinction between climate change attributed to human activities changing the atmospheric composition and climate variability attributed to natural causes. Scientific evidence has shown that climate change might be caused by internal forcing processes (i.e., continental drift) or external forcings such as modulations of the solar cycles, volcanic eruptions and persistent anthropogenic changes in atmosphere composition or in land use (IPCC 2018).

Profound changes in climate are underway on our planet. A key aspect and an essential driver of climate change is global warming. The IPCC defines global warming as "*the estimated increase in global mean surface temperature (GMST) averaged over a 30 year period, or the 30 year period centered on a particular year or decade, expressed relative to pre-industrial levels unless otherwise specified*" (IPCC 2018). Since the middle of nineteenth century, global surface temperature has

K. Tang (✉)
School of Economics and Trade, Guangdong University of Foreign Studies, Guangzhou 510006, China
e-mail: francistang1988@hotmail.com

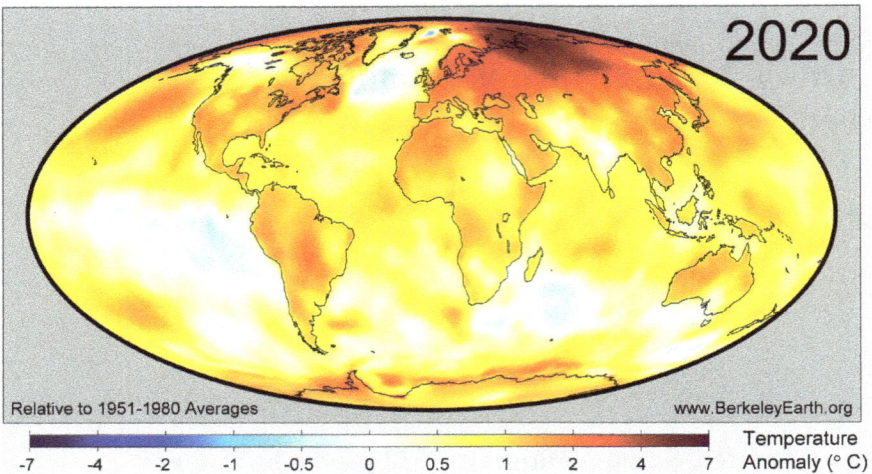

Fig. 1.1 Temperatures in 2020 relative to the average temperature in 1951–1980. *Source* Berkeley Earth (2021)

risen by more than 1 degree Celsius (°C). Two-thirds of the warming has occurred since 1975, with a rate of about 0.15–0.20 °C each ten years (Fig. 1.1).

Over the last half century, observed global mean surface temperature has risen at an unprecedented rate in the last 20 centuries (Fig. 1.2). Global surface temperature was 1.09 °C higher in the latest decade (2011–2020) than the average temperature of the late nineteenth century, with higher increases over land than over the ocean (Fig. 1.3) (IPCC 2021). Globally, 2016 was the warmest year since 1850, followed by 2020. The global average surface temperature in 2016 was 1.29 °C higher than the average of 1850–1900, and about 0.02 °C higher than 2020. 2020 had a new record for the warmest annual land-average (Berkeley Earth 2021).

China is a sensitive area of global climate change. In China, observed mean surface temperature has risen 1.15 °C over the last century, with a rate of about 0.10 °C per decade. Since the middle of the twentieth century, the country's warming rate has become higher (Fig. 1.4).

According to the Blue Paper of Climate Change in China (2021) issued by the China National Climate Centre, the country's warming rate, which was roughly 0.26 °C per decade for the 1951–2020 period, is "significantly higher" than the global average. The Qinghai-Tibet plateau had the highest warming rate among all regions in the country, with an average of 0.37 °C per decade. The latest two decades (2001–2020) was the warmest period in China since 1901. Nine of the ten warmest years since 1901 have been observed in the twenty-first century.

It is highly likely that the widespread shrinking of the global cryosphere, a result of global warming, is causing sea level rise. Melting glaciers and ice sheets, reducing snow cover, and massive loss from sea ice are adding water to the ocean. Since the beginning of the twentieth century, global mean sea level has risen about 200 mm,

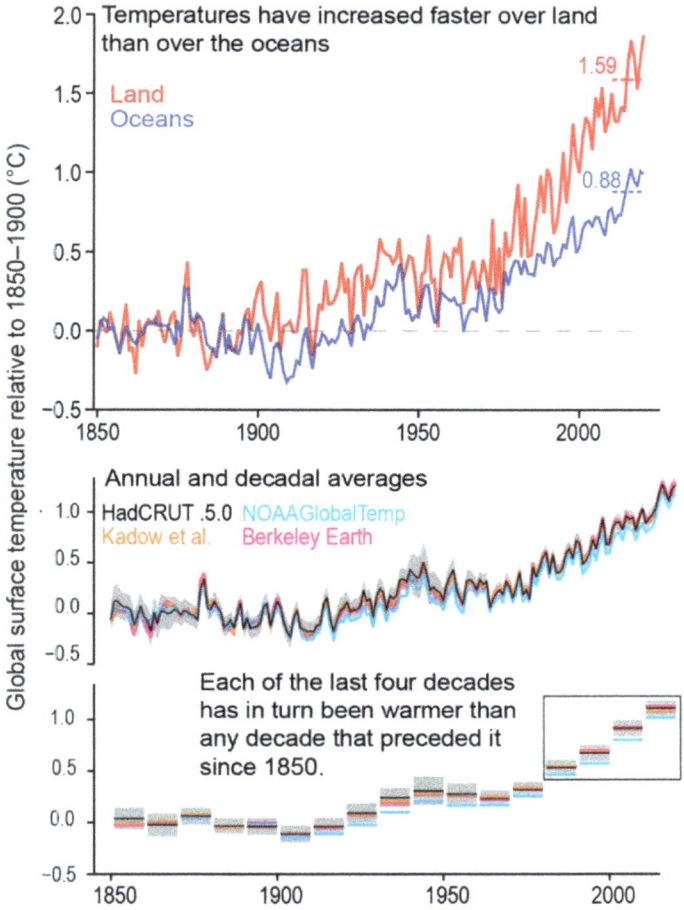

Fig. 1.2 Global surface temperature relative to the average temperature in 1850–1900. *Source* IPCC (2021)

more than in any century in last 3000 years. The rising rate has accelerated over the last half century. The annual average rates for the 1971–2006 and 2006–2018 periods were 1.87 mm and 3.69 mm, respectively. Scientific evidence confirms that such a rising trend is going to continue for the coming centuries. IPCC (2019) projected 600–1100 mm of rise by 2100 with high GHG emissions, while 300–600 mm of rise by 2100 with substantial emissions cuts (Fig. 1.5). IPCC (2021) projected that the rise would likely be 380–770 mm by 2100, compared to the 1995–2014 level.

China's coastal sea waters have continued to rise at an accelerated rate in the latest four decades (Fig. 1.6). On average, the country's sea levels have risen at an annual rate of 3.4 mm, which is higher than the global average, for the 1980–2020 period. In 2020, China's coastal waters were 73 mm higher than the 1993–2011 level, which hit

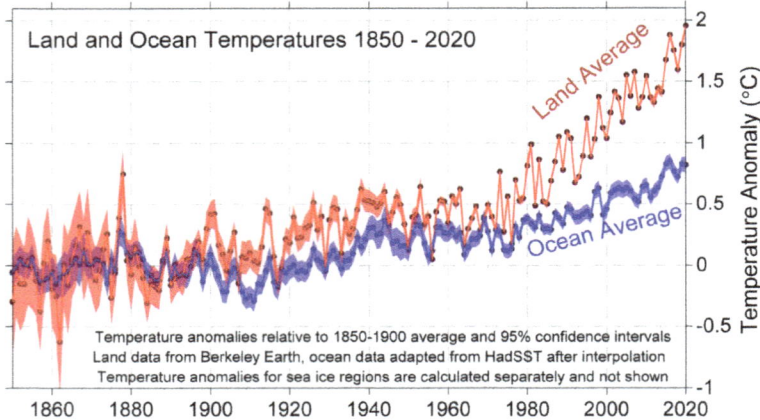

Fig. 1.3 Land and ocean temperature changes relative to the average temperature in 1850–1900. *Source* Berkeley Earth (2021)

the 3rd highest level since 1980 and were 25 mm higher than the 2018 level (Fig. 1.7). Coastal waters in the Bohai, Yellow, East China, and South China Seas were 86, 60, 79, and 68 mm higher than the 1993–2011 levels, respectively. It is projected that the country's sea levels would likely rise by 55–170 mm in the next three decades.

Climate change has worsened the frequency and intensity of extreme weather events since the middle of twentieth century. Heatwaves, droughts, floods and storms are likely to become more frequent and intense with warming climate, even if global warming is held to 1.5 °C above pre-industrial levels. Since the 1950s, heatwaves have become longer, hotter and more common in North America, Europe, Asia and Australia. Severe droughts become more frequent in the Mediterranean region, Central Europe, Middle America, Southern Australia, and Eastern and Southern Africa. Since 1950, floods have increased over most areas of Eastern and Southern Asia and North America. Climate change is also increasing the threat from storms in southern China, Southeaster and Southern Asia.

China has also seen more extreme weather events like heavy precipitation and heatwaves. The Blue Paper of Climate Change in China (2021) confirmed increasing extreme precipitation events since the 1960s (Fig. 1.8) and rising heatwaves and tropical storms since the 1990s. More frequent and intensive extreme weather events in China have wreaked havoc on livelihoods and local economies, and caused tens of billions of dollars in damage every year (China National Climate Centre 2021).

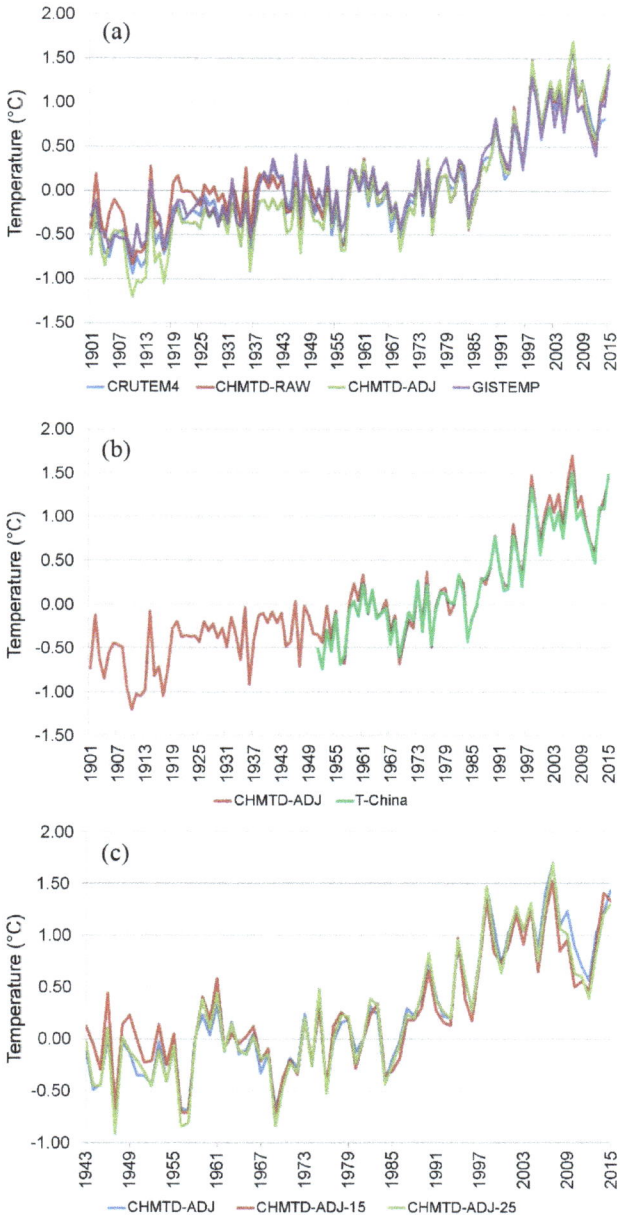

Fig. 1.4 a The mean temperature anomaly series for China during 1901–2015 based on the original (CHMTD-RAW) and adjusted (CHMTD-ADJ) station data, compared with those based on CRUTEM4 and GISTEMP. The anomalies are from the 1961–1990 climatology. **b** The CHMTD–ADJ series (1901–2015) versus the T-China series (1951–2015) based on 2419 stations. **c** The mean temperature anomaly series for China during 1943–2015. *Source* Cao et al. (2017)

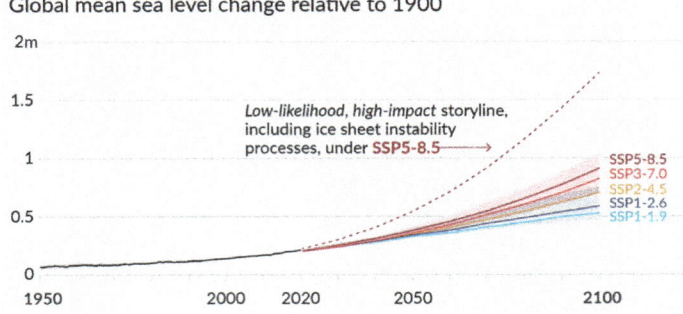

Fig. 1.5 Global average sea level change. *Source* IPCC (2021)

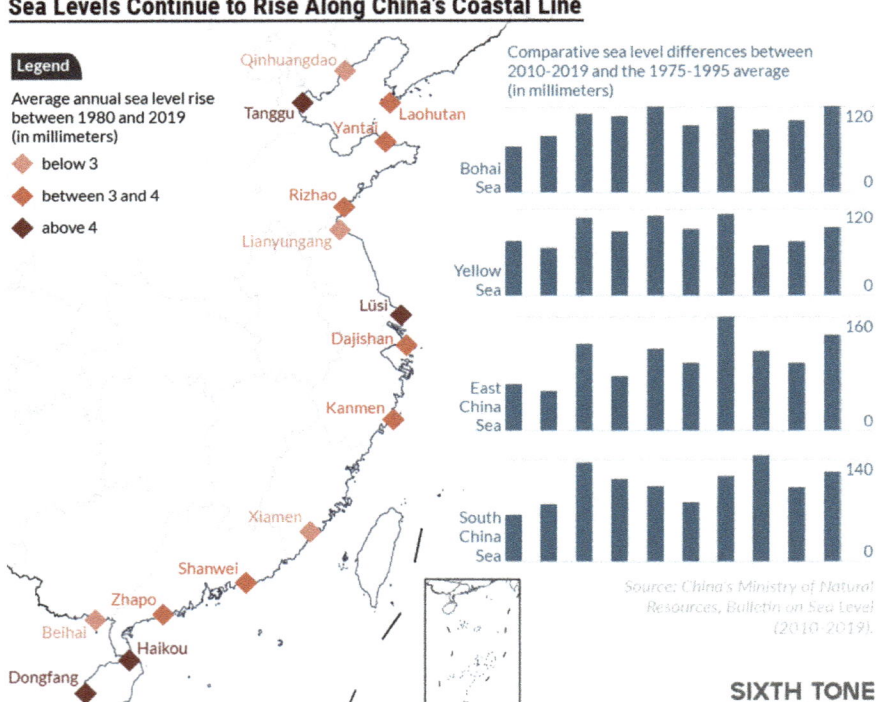

Fig. 1.6 Sea levels in China. *Source* Six Tone (2020).[1] Data source: Ministry of Natural Resources (2010–2019)

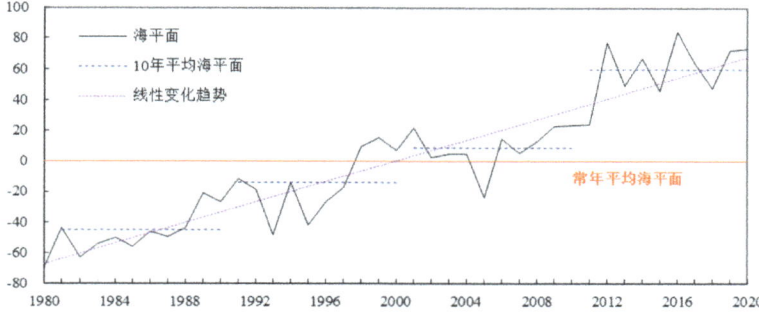

Fig. 1.7 Coastal sea level change in China, 1980–2020 (mm). The line in the middle of the figure shows the 1993–2011 level. *Source* Ministry of Natural Resources (2021)

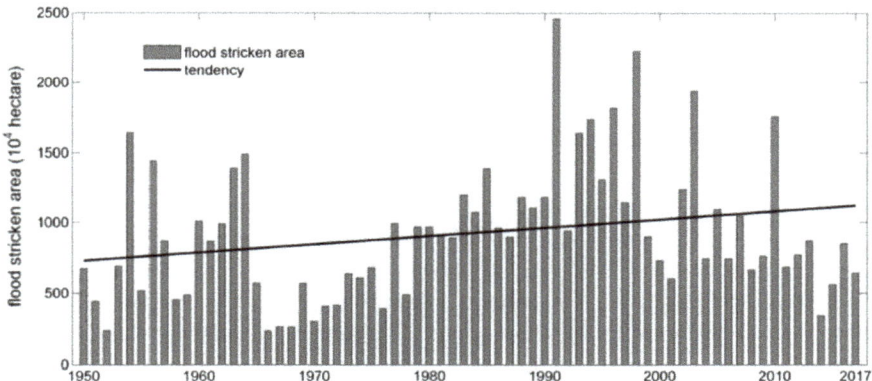

Fig. 1.8 Flood stricken area in China, 1950–2017. *Source* Kundzewicz et al. (2019)

1.2 Anthropogenic Greenhouse Gas Emissions: The Primary Driver of Climate Change

Though there are debates on the causes of climate change, the vast majority of the global scientific community agrees that anthropogenic emissions of GHG, such as carbon dioxide (CO_2), methane (CH_4) and nitrous oxide (N_2O), are the primary driver of the changing climate. GHG emissions from human activities have substantially increased the concentration of GHG in the atmosphere since the 1750s, thus increasing the greenhouse effect and global warming. IPCC (2021) confirmed that anthropogenic GHG emissions are responsible for roughly 1.1 °C of warming between 1850–1900 and 2010–2019.

CO_2 is the primary GHG emitted through human activities and contributes about three-quarters of total GHG emissions. CO_2 emissions are mainly from fossil fuels

---

[1] https://image5.sixthtone.com/image/5/33/906.jpg.

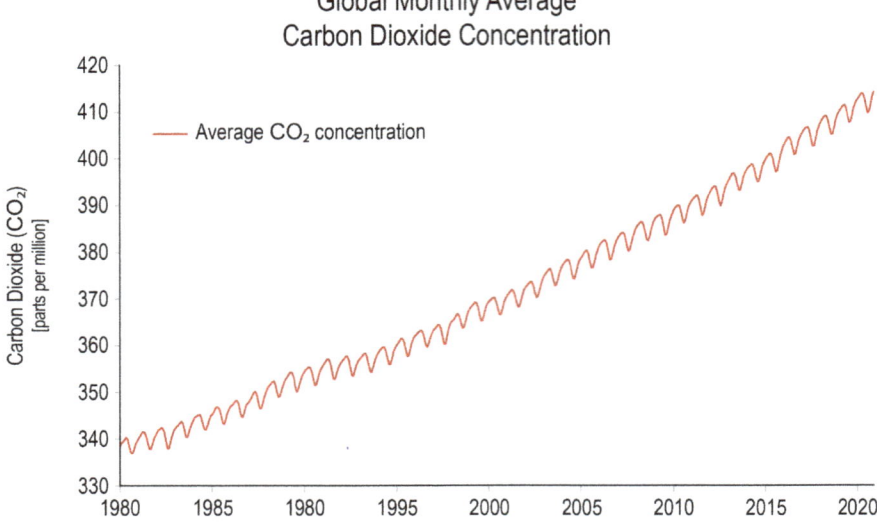

Fig. 1.9 Global monthly average CO_2 concentration, 1980–2020. *Source* Global Change (2021)[2]

burning (e.g., coal, oil and natural gas). Burning them is responsible for around 90% of human-related CO_2 emissions (Welsby et al. 2021). Global atmospheric concentrations of CO_2 in 2011, 2017 and 2018 were 391 ppm, 405 ppm and 407.8 ppm, respectively. In May 2021, atmospheric concentrations of CO_2 reached 419 ppm, higher than at any time in 2 million years. The annual increasing rates of atmospheric CO_2 concentrations for the 1985–1995, 1995–2005 and 2005–2015 periods were 1.42 ppm, 1.86 ppm and 2.06 ppm, respectively (Fig. 1.9). CO_2 had increased by more than 47% from 1750 to 2019 (IPCC 2021).

CH_4 is the second largest anthropogenic GHG after CO_2, contributing about one fifth of the world's GHG emissions. About 60% CH_4 emissions are from human activities, such as livestock and rice production and natural gas exploitation. In 2019, concentrations of CH_4 were 1866.3 ppb, higher than at any time in 800,000 years. The concentrations further increased and reached 1891.3 ppb in April 2021 (Fig. 1.10). IPCC (2021) shows that the increase since 1750 *"far exceeds the range over multiple glacial-interglacial transitions of the past 800, 000 years"*. Since 1800, CH_4 concentrations in the atmosphere have more than doubled, largely due to human-related activities. The average increase rate between 2010 and 2019 was about 7.6 ppb per year.

N_2O is also a potent GHG contributing to human-induced climate change (Montzka et al. 2011). Human activities such as agriculture, wastewater management, and industrial processes are contributing to the amount of N_2O in the atmosphere. More than 70% of those N_2O emissions come from agriculture. IPCC (2018) estimated that is the thrid largest anthropogenic GHG, contributing about 6% of

---

[2] https://www.globalchange.gov/browse/indicators/atmospheric-carbon-dioxide.

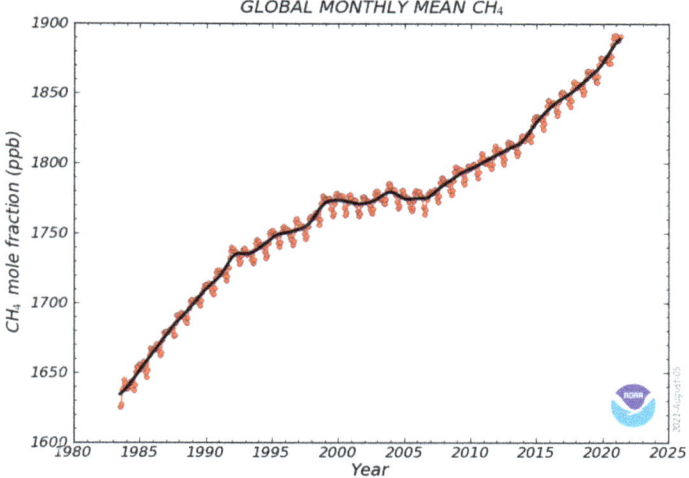

Fig. 1.10 Global monthly mean CH₄ concentration. *Source* NOAA/GML (2021)[3]

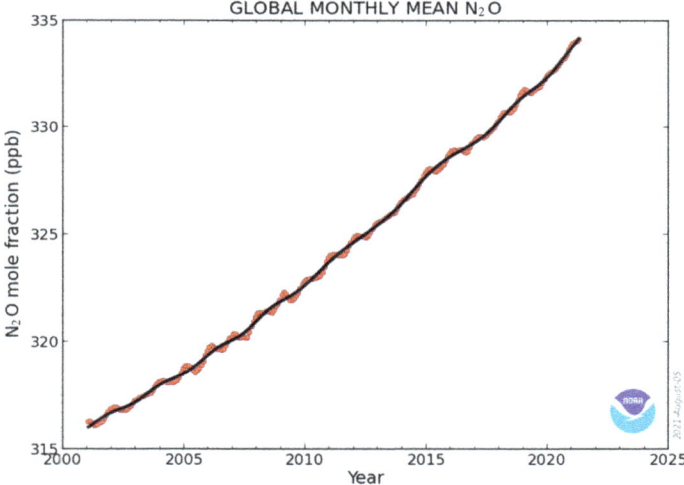

Fig. 1.11 Global monthly mean N₂O concentration. *Source* NOAA/GML (2021)[4]

the global GHG emissions. In 2019, concentrations of N_2O were 332.1 ppb, higher than at any time in 800,000 years. The concentrations further increased and reached

334.1 ppb in April 2021 (Fig. 1.11). In the latest four decades, increased N_2O emissions have been largely driven by the expansion of agricultural activities. N_2O had increased by about 23% from 1750 to 2019 (IPCC 2021).

The total warming effect from anthropogenic GHG had increased by 45% for the 1990–2019 period. The warming effect from CO_2 had increased by 36%. Compared to the 1950–1900 level, global surface temperature would be very likely higher by up to 5.7 °C with high GHG emissions (IPCC 2021). Enormous scientific evidence argues that stabilising the earth temperature rise at 1.5 °C would contribute to avoiding the most serious effects of changing climate. Thus, it is vital to substantially cut GHG emissions as soon as possible.

1.3   Net-Zero Emissions: The world's Critical Mission

Since the 1990s, the global community has taken measures to combat climate change. In 1992, most of the United Nations members signed the United Nations Framework Convention on Climate Change (UNFCCC), calling for limiting human GHG emissions. As of 2020, the UNFCCC has 197 signatory parties. The Kyoto Protocol, signed in 1997 for the 2005–2020 period, was the first implementation of actions under the UNFCCC. The Protocol required 37 industrialized countries and the EU to limit and reduce their GHG emissions with individual targets. The Protocol was superseded by the Paris Agreement entered into force in 2016. The Agreement, as showed in its Article 2, aims to *"strengthen the global response to the threat of climate change, in the context of sustainable development and efforts to eradicate poverty, including by: (a) Holding the increase in the global average temperature to well below 2 °C above pre-industrial levels and to pursue efforts to limit the temperature increase to 1.5 °C above pre-industrial levels, recognizing that this would significantly reduce the risks and impacts of climate change; (b) Increasing the ability to adapt to the adverse impacts of climate change and foster climate resilience and low greenhouse gas emissions development, in a manner that does not threaten food production; (c) Making finance flows consistent with a pathway towards low greenhouse gas emissions and climateresilient development."*

The key goal of the Paris Agreement is to limit global warming to well below 2 °C and pursue efforts to limit the temperature increase to 1.5 °C above pre-industrial levels. This requires the global community to achieve net-zero emissions, an essential and the only way to halt serious climate change, around the middle of the twenty-first century, or as soon as possible. Net-zero emissions mean achieving a balance between GHG emitted and removed. The IPCC defines net-zero emissions as *"when anthropogenic emissions of greenhouse gases to the atmosphere are balanced by anthropogenic removals over a specified period"* (IPCC 2018). The UN Climate Change defines net-zero emissions as *"a state where a balance between anthropogenic GHG*

---

[3] https://gml.noaa.gov/ccgg/trends_ch4/

[4] https://gml.noaa.gov/ccgg/trends_n2o/

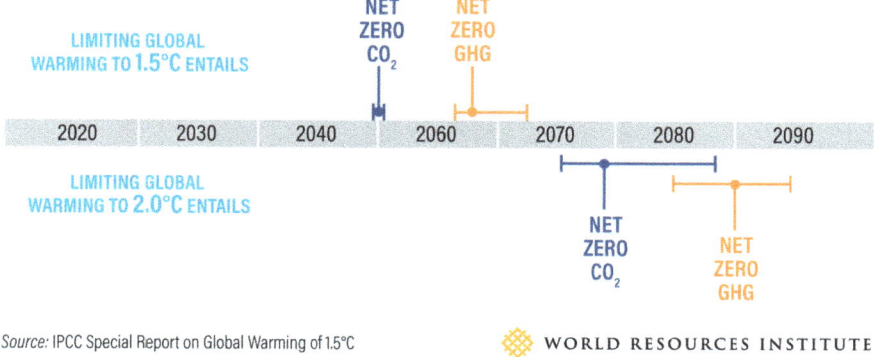

Fig. 1.12 Global net-zero emissions timeline. *Source* IPCC (2018)

emissions and removals is achieved" (UNCC 2021). Noticeably, carbon neutrality means achieving net-zero CO_2 emissions. In China, the term carbon neutrality is widely used to refer to net-zero emissions by media and politicians. It should be noted that net-zero emissions actually deliver a more comprehensive and broader commitment to tackle all GHG emissions. IPCC (2018) provided the global timeline to reach net-zero emissions for the 1.5 and 2 °C goals (Fig. 1.12).

1.4 China's Emissions and Net-Zero Target

China has been the largest emitter of GHG in the world for more than one decade. During the latest several decades, the country's GHG emissions have increased rapidly accompanied by its continuously expanding population and booming economy (Fig. 1.13). In 2006, China overtook the US as the global largest CO_2 emitter. Every year, China emits more GHG than the entire developed countries combined. In 2019, China emitted 27% of the world's GHG, while the US and India contributed 11% and 6%, respectively. GHG emissions from all members of the Organization for Economic Cooperation and Development (OECD) as well as all 27 EU member states were 14,057 MMt CO_2e, while emissions in China were 14,093 MMt CO_2 (Larsen et al. 2021). Over the past decade, China has accounted for more than 60 per cent of global emissions growth (IEA 2018). Because of its 1.4 billion population, China's per capita emissions, 10.1 tons, were still lower than the OECD level (10.5 tons). Obviously, the global fight against climate change would eventually fail without China's efforts (Table 1.1).

In China, coal, one of the most polluting forms of energy, has been the dominant source of energy for decades, and its use is still increasing (Fig. 1.14). In 2020, the share of coal in primary energy consumption mix was 56.8%, while the share was 72.4% in 2005 and roughly 70% in 2010. The dominant coal was followed by oil

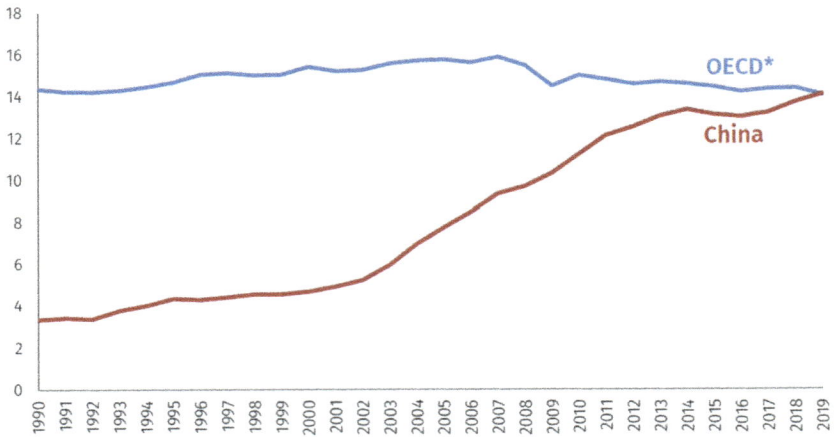

Source: Rhodium Group, UNFCCC. Includes emissions and removals of land-use, land-use change and forests (LULUCF). Excludes international aviation and marine bunkers. Includes six Kyoto gases using AR4 GWP values. *OECD includes OECD members as of 2019 and includes all EU member states.

Fig. 1.13 Net greenhouse gas emissions (Gigatons of CO_2e). *Source* Larsen et al. (2021)

(20%), clean electricity (hydro, wind, solar and nuclear) (9%), natural gas (8%), and biomass (3%) (Department of Energy Statistics 2021). Globally, China accounts for more than half of the world's total coal use.[5] In 2020, the country's coal generation rose by 1.7%, and energy consumption still grew, despite the COVID-19 outbreak, but at a slower rate of 2.2%. Total energy consumption had been growing slightly faster since 2017 (3.3% annually compared to 1.7% annually over 2012–2017).[6]

Under both international and domestic pressures to do more to address global warming, China's leaders have recently pledged to peak its emissions within the next ten years, and become carbon neutral before 2060. In September 2020, President Xi Jinping declared that China aims to peak CO_2 emissions before 2030 and achieve carbon neutrality by 2060. This is the first long-term climate goal of the most-populous country, which could affect more than the country's 1.4 billion people. Considering China's ongoing rapid industrialization and urbanization and particular socioeconomic and political realities, the goal is challenging. In addition, China's current administration has recognized climate change as its top priority. In response, more 2030 pledges have been made, including increasing the share of non-fossil fuels in primary energy consumption mix to 25%, decreasing carbon intensity, the amount of CO_2 emissions per unit of GDP, by 65%, increasing the combined capacity of solar and winder power generators to 1.2 billion kilowatts, and enhancing the forest stock volume by 6 billion cubic meters from the 2005 level.[7] In addition, China has

---

[5] See https://ember-climate.org/global-electricity-review-2021/g20-profiles/china/ for more information.

[6] https://www.enerdata.net/publications/daily-energy-news/coal-may-account-44-chinas-energy-mix-2030.html.

[7] http://www.xinhuanet.com/politics/leaders/2020-12/12/c_1126853600.htm.

Table 1.1 International net-zero emissions goals (Until August 2021)

Country/Region	Time	Status	Country/Region	Time	Status	Country/Region	Time	Status
Suriname	–	Achieved	Finland	2035	In policy	Nepal	2050	In policy
Bhutan	–	Achieved	Austria	2040	In policy	Laos	2050	In policy
Germany	2045	In law	Iceland	2040	In policy	Jamaica	2050	In policy
Sweden	2045	In law	US	2050	In policy	Mauritius	2050	In policy
EU	2050	In law	South Africa	2050	In policy	Monaco	2050	In policy
Japan	2050	In law	Italy	2050	In policy	Malawi	2050	In policy
UK	2050	In law	Brazil	2050	In policy	Maldives	2050	In policy
France	2050	In law	Switzerland	2050	In policy	Barbados	2050	In policy
Denmark	2050	In law	Argentina	2050	In policy	Andorra	2050	In policy
Spain	2050	In law	Norway	2050	In policy	Cabo Verde	2050	In policy
Hungary	2050	In law	Colombia	2050	In policy	Grenada	2050	In policy
Luxembourg	2050	In law	Portugal	2050	In policy	Vatican City	2050	In policy
Canada	2050	In law	Slovakia	2050	In policy	Marshall Islands	2050	In policy
New Zealand	2050	In law	Panama	2050	In policy	Nauru	2050	In policy
South Korea	2050	Proposed legislation	Dominican Republic	2050	In policy	China	2060	In policy
Ireland	2050	Proposed legislation	Costa Rica	2050	In policy	Indonesia	2060	In policy
Chile	2050	Proposed legislation	Uruguay	2050	In policy	Kazakhstan	2060	In policy
Fiji	2050	Proposed legislation	Slovenia	2050	In policy	Ukraine	2060	In policy
			Latvia	2050	In policy	India	2070	In policy

formulated and implemented a variety of strategies, regulations, policies, standards, and actions (Wu et al. 2021a, b; Yuan et al. 2021, 2022; Chen et al. 2022).

China has taken measures to combat climate change. The country's policymakers are improving overall planning and coordination in response to climate change and accelerating work on 1 + N policies for peaking carbon emissions and achieving

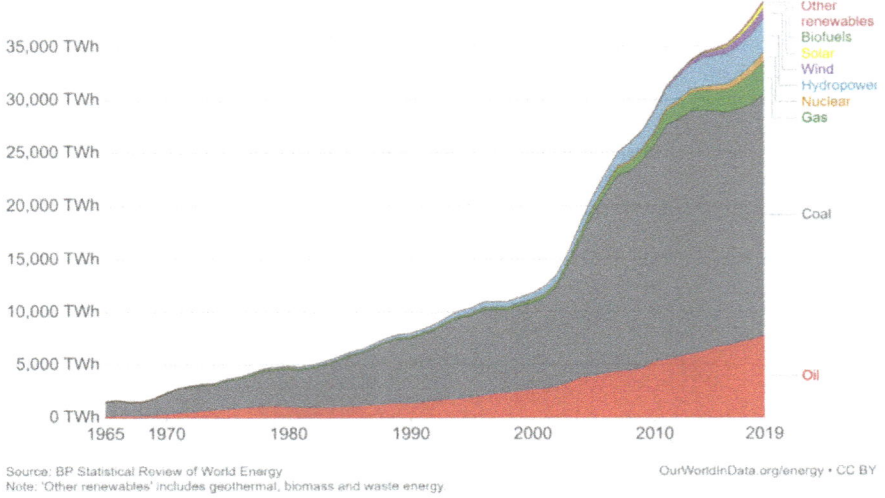

Fig. 1.14 China's energy mix, 1965–2019. *Source* Araral (2021)

carbon neutrality. The response to climate change has been incorporated into national economic and social development plans. A mechanism of breaking down and meeting the climate targets has been established (The State Council Information Office 2021). Moreover, multiple strategies, such as enhancing energy efficiency, optimizing energy mix and adjusting industrial structure, have been adopted to slow the growth of the country's GHG emissions. Market-based mechanism has also been introduced (Tang et al. 2019; Tang and Hailu 2020). Since 2013, eight pilot carbon trading markets have been established, covering more than 11% of the country's carbon emissions (Tang et al. 2021a, b; Tang and Ma 2022; Zhang et al. 2017). The established ETS schemes, which are stand-alone regional markets with differing designs, cover selected local major emitting industries. A national carbon trading market, which is the world's biggest, was officially launched in July 2021. The newly established market covers 2,225 domestic power operators using fossil fuels, accounting for more than 14% of fossil-fuel-related CO_2 emissions in the world. As a result, the country's carbon intensity in 2020 was 18.8% lower than the 2015 level, 24.4% lower than 2012, and more than 48% lower than 2005. The share of non-fossil fuels in primary energy consumption mix reached 15.3% in 2019, ahead of the planned target of 15% by 2020. The country has met nine of the fifteen quantitative targets in its 2015 climate commitments ahead of schedule (The State Council Information Office 2021).

However, for the international community and China itself, reining in CO_2 emissions is not enough to tackle climate change. The announced goals are not ambitious enough. The long-term net-zero goal is yet to be matched with short-term plans. Major changes are needed to secure net-zero emissions by 2060. Despite China's efforts to push for carbon neutrality, its net-zero pathways are yet not quite clear, especially from an economic perspective. It seems that China is combating climate change continuing the progress already being achieved. The country has not accelerated its pace of decarbonization, partly due to the concern of the possible associated negative impacts on economic growth and consumption in a large-scale and high-complex developing economy. If China do not peak its emissions quickly, achieving net-zero emissions by the middle of this century is going to be a real challenge for the world.

Scientific studies suggest that a balanced mixture of strategies aiming to reduce and offset GHG emissions is needed to achieve carbon neutrality. Establishing net-zero pathways needs to identify the best combination of these strategies by considering their costs and benefits, tradeoffs, interdependence, and impacts to economic development, social welfare, and environmental quality. Furthermore, net-zero pathways need to be customized for different actors ranging from producers, industries, regions, and the country. Net-zero goals and pathways of these actors might contradict each other. China, together with the international community, should and must do more on addressing climate change.

1.5 Conclusions

This chapter addresses why China's net-zero emissions matter to us. Profound changes in climate are underway on our planet. Since the middle of nineteenth century, global surface temperature has risen by more than 1 °C. Since the beginning of the twentieth century, global mean sea level has risen about 200 mm. Extreme weather events have become more frequent and intensive since the middle of twentieth century. GHG emissions from human activities have substantially increased the concentration of GHG in the atmosphere since the 1750s, thus increasing the greenhouse effect and global warming. The key goal of the Paris Agreement is to limit global warming to well below 2 °C and pursue efforts to limit the temperature increase to 1.5 °C above pre-industrial levels. This requires the global community to achieve net-zero emissions soon. Since China is the world's largest emitter of GHG, the global fight against climate change would eventually fail without China's net-zero efforts. The world needs China to succeed.

However, for the international community and China itself, reining in CO_2 emissions is not enough to tackle climate change. Major changes are needed to secure net-zero emissions by 2060. Despite China's efforts to push for carbon neutrality, its net-zero pathways are yet not clear. The country has not accelerated its pace of decarbonization. If China do not peak its emissions quickly, achieving net-zero emissions by the middle of this century is going to be a real challenge for the world.

References

Araral, E. (2021). What can the G-7 learn from China's transition to climate-smart growth?. Brookings. https://www.brookings.edu/blog/future-development/2021/06/16/what-can-the-g-7-learn-from-chinas-transition-to-climate-smart-growth/

Berkeley Earth (2021). Global Temperature Report for 2020. http://berkeleyearth.org/global-temperature-report-for-2020/

Cao, L., Yan, Z., Zhao, P., Zhu, Y., Yu, Y., Tang, G., & Jones, P. (2017). Climatic warming in China during 1901–2015 based on an extended dataset of instrumental temperature records. *Environmental Research Letters*, 12(6), 064005.

Chen, X., Chen, G., Lin, M., Tang, K., & Ye, B. (2022). How does anti-corruption affect enterprise green innovation in China's energy-intensive industries?. *Environmental Geochemistry and Health*, 44(9), 2919–2942. https://doi.org/10.1007/s10653-021-01125-4

China National Climate Centre (2021). *Blue Paper of Climate Change in China (2021)*. Science Press, Beijing.

Department of Energy Statistics, National Bureau of Statistics (2021). *China Energy Statistical Yearbook 2021*. China Statistics Press, Beijing.

IPCC (2018). *Global Warming of 1.5°C. An IPCC Special Report on the impacts of global warming of 1.5°C above pre-industrial levels and related global greenhouse gas emission pathways, in the context of strengthening the global response to the threat of climate change, sustainable development, and efforts to eradicate poverty* [Masson-Delmotte, V., P. Zhai, H.-O. Pörtner, D. Roberts, J. Skea, P.R. Shukla, A. Pirani, W. Moufouma-Okia, C. Péan, R. Pidcock, S. Connors, J.B.R. Matthews, Y. Chen, X. Zhou, M.I. Gomis, E. Lonnoy, T. Maycock, M. Tignor, and T. Waterfield (eds.)]. In Press.

Fankhauser, S., Smith, S. M., Allen, M., et al. (2022). The meaning of net zero and how to get it right. *Nature Climate Change*, 12(1), 15–21.

IEA. (2018). *CO_2 emissions from fuel combustion 2018*. Paris: International Energy Agency.

IPCC (2019). *IPCC Special Report on the Ocean and Cryosphere in a Changing Climate* [H.-O. Pörtner, D.C. Roberts, V. Masson-Delmotte, P. Zhai, M. Tignor, E. Poloczanska, K. Mintenbeck, A. Alegría, M. Nicolai, A. Okem, J. Petzold, B. Rama, N.M. Weyer (eds.)]. In press.

IPCC (2021). *Climate Change 2021: The Physical Science Basis. Contribution of Working Group I to the Sixth Assessment Report of the Intergovernmental Panel on Climate Change* [Masson-Delmotte, V., P. Zhai, A. Pirani, S. L. Connors, C. Péan, S. Berger, N. Caud, Y. Chen, L. Goldfarb, M. I. Gomis, M. Huang, K. Leitzell, E. Lonnoy, J. B. R. Matthews, T. K. Maycock, T. Waterfield, O. Yelekçi, R. Yu and B. Zhou (eds.)]. Cambridge University Press. In Press.

Kundzewicz, Z. W., Su, B., Wang, Y., Xia, J., Huang, J., & Jiang, T. (2019). Flood risk and its reduction in China. *Advances in Water Resources*, 130, 37–45.

Larsen, K., Pitt, H., Grant, M., & Houser, T. (2021). China's greenhouse gas emissions exceeded the developed world for the first time in 2019. Rhodium Group. https://rhg.com/research/chinas-emissions-surpass-developed-countries/#_ftn1

Ministry of Natural Resources (2010–2019). Bulletin on Sea Level. Ministry of Natural Resources, Beijing.

Ministry of Natural Resources (2021). Bulletin on Sea Level. Ministry of Natural Resources, Beijing.

Montzka, S. A., Dlugokencky, E. J., & Butler, J. H. (2011). Non-CO_2 greenhouse gases and climate change. *Nature*, 476(7358), 43–50.

Tang, K., & Hailu, A. (2020). Smallholder farms' adaptation to the impacts of climate change: Evidence from China's Loess Plateau. *Land Use Policy*, 91, 104353. https://doi.org/10.1016/j.landusepol.2019.104353

Tang, K., & Ma, C. (2022). The cost-effectiveness of agricultural greenhouse gas reduction under diverse carbon policies in China. *China Agricultural Economic Review*. https://doi.org/10.1108/CAER-01-2022-0008

Tang, K., He, C., Ma, C., & Wang, D. (2019). Does carbon farming provide a cost-effective option to mitigate GHG emissions? Evidence from China. *Australian Journal of Agricultural and Resource Economics*, 63(3), 575–592. https://doi.org/10.1111/1467-8489.12306

Tang, K., Liu, Y., Zhou, D., & Qiu, Y. (2021a). Urban carbon emission intensity under emission trading system in a developing economy: Evidence from 273 Chinese cities. *Environmental Science and Pollution Research*, 28(5), 5168–5179. https://doi.org/10.1007/s11356-020-10785-1

Tang, K., Zhou, Y., Liang, X., & Zhou, D. (2021b). The effectiveness and heterogeneity of carbon emissions trading scheme in China. *Environmental Science and Pollution Research*, 28(14), 17306–17318. https://doi.org/10.1007/s11356-020-12182-0

The State Council Information Office (2021). *Responding to Climate Change: China's Policies and Actions*. Beijing: The State Council Information Office of the People's Republic of China. http://www.scio.gov.cn/zfbps/32832/Document/1715506/1715506.htm#:~:text=China%20has%20announced%20in%202021,emissions%20and%20achieving%20carbon%20neutrality

UN Climate Change (UNCC) (2021). Climate Neutral Now: Guidelines for Participation. https://unfccc.int/documents/271233

UNFCCC (2011). Fact sheet: Climate change science. https://unfccc.int/files/press/backgrounders/application/pdf/press_factsh_science.pdf

Welsby, D., Price, J., Pye, S., & Ekins, P. (2021). Unextractable fossil fuels in a 1.5° C world. *Nature*, 597(7875), 230–234.

Wu, J., Feng, Z., & Tang, K. (2021a). The dynamics and drivers of environmental performance in Chinese cities: A decomposition analysis. *Environmental Science and Pollution Research*, 28(24), 30626–30641. https://doi.org/10.1007/s11356-021-12786-0

Wu, J., Xu, H., & Tang, K. (2021b). Industrial agglomeration CO_2 emissions and regional development programs: A decomposition analysis based on 286 Chinese cities. *Energy*, 225, 120239. https://doi.org/10.1016/j.energy.2021.120239

Yuan, F., Tang, K., & Shi, Q. (2021). Does Internet use reduce chemical fertilizer use? Evidence from rural households in China. *Environmental Science and Pollution Research*, 28(5), 6005–6017. https://doi.org/10.1007/s11356-020-10944-4

Yuan, F., Tang, K., Shi, Q., Qiu, W., & Wang, M. (2022). Rural women and chemical fertiliser use in rural China. *Journal of Cleaner Production*, 344, 130959. https://doi.org/10.1016/j.jclepro.2022.130959

Zhang, J., Wang, Z., & Du, X. (2017). Lessons learned from China's regional carbon market pilots. *Economics of Energy & Environmental Policy*, 6(2), 19–38.

Chapter 2
Spatiotemporal Dynamics of China's Carbon Emissions: Evidence from Urban Efficiency and Convergence

Kai Tang and Di Zhou

2.1 Introduction

Since 1980s, China's economy has been growing remarkably. However, the accompanied rapid urbanisation and industrialisation have driven fast increase in energy consumption, thus boosting carbon emissions. In 2018, China produced about 16% of global gross domestic product (GDP) and generated more than 29% of global greenhouse gases (GHG) emissions (mainly CO_2 emissions).[1] Currently, the country is struggling with limiting enormous CO_2 emissions and maintaining relatively high economic growth speed concurrently, which has been reflected by its recent ambitious national goals.

The Chinese government released a document titled *Working Guidance for Carbon Dioxide Peaking and Carbon Neutrality in Full and Faithful Implementation of the New Development Philosophy* on 22 September 2021.[2] This document includes several updated national goals in terms of limiting carbon emissions and net-zero emissions: (1) *By 2025, energy consumption per unit of GDP will be lowered by 13.5% from the 2020 level; CO_2 emissions per unit of GDP will be lowered by 18% from the 2020 level; the share of non-fossil energy consumption will have reached around 20%; the forest coverage rate will have reached 24.1%, and the forest stock volume will have risen to 18 billion cubic meters.* (2) *By 2030, energy consumption per unit of GDP will have declined significantly; CO_2 emissions per unit of GDP will have dropped by more than 65% compared with the 2005 level; the share of non-fossil energy consumption will have reached around 25%, with the total installed*

---

[1] https://www.imf.org/external/pubs/ft/weo/2019/01/weodata/index.aspx; https://www.bp.com/en/global/corporate/energy-economics/statistical-review-of-world-energy.html.

[2] http://www.gov.cn/zhengce/2021-10/24/content_5644613.htm.

K. Tang (✉) · D. Zhou
School of Economics and Trade, Guangdong University of Foreign Studies, Guangzhou 510006, China
e-mail: francistang1988@hotmail.com

capacity of wind power and solar power reaching over 1200 gigawatts; the forest coverage rate will have reached about 25%, and the forest stock volume will have reached 19 billion cubic meters. CO_2 emissions will reach peak and stabilization and then decline. (3) By 2060, China will have fully established a green, low-carbon and circular economy and a clean, low-carbon, safe and efficient energy system. Energy efficiency will be at the advanced international level, and the share of non-fossil energy consumption will be over 80%. China will be carbon neutral, and it will have achieved fruitful results in ecological civilization and reached a new level of harmony between humanity and nature.[3]

It is generally agreed by existing literature that enhancing carbon emissions efficiency (CEE) plays an essential role in promoting net-zero emissions and improving the sustainability of socioeconomic systems (Tang et al. 2016, 2021; Yang et al. 2018; Iram et al. 2020; Tang and Hailu 2020). In recent years, there is an increasing number of studies using diverse methods to analyse CEE in multiple countries/regions (e.g., Mirza and Kanwal 2017; Tang et al. 2019; Chang 2020). Overall, those studies recognise that substantial spatial heterogeneity of CEE among countries exists, which may have negative impact on regional and/or global net-zero emissions progress. Therefore, it is necessary to know whether low CEE countries could catch up with high CEE ones, which is also the key topic of global CEE convergence research.

Considering that traditional convergence approaches, including β convergence, σ convergence, and random convergence tests (e.g., Barro and Sala-I-Martin 1992; Liu et al. 2017), ignore individual heterogeneity, Barro and Sala-I-Martin (1991) and Quah (1997) demonstrated the club convergence, which describes the convergence in the economic growth of regions sharing similar initial socioeconomic conditions. Evaluating the club convergence of CEE, especially within a country, could show the movements and drivers of regional CEE in differing convergence club regions (Tang et al. 2021), which could also deliver empirical insights and policy reference for coordinating net-emissions progress among regional socioeconomic systems (Haider and Akram 2019). Nevertheless, relevant empirical evidence is rather rare.

Some recent studies focusing on China have attempted to evaluate the club convergence of CEE employing provincial or sectoral data (Yang et al. 2017; Lin and Wang 2019; Wang et al. 2019a; Zhou et al. 2020). However, few have addressed this issue based on city-level data. China's cities emit about 85% of the country's total CO_2 emissions (Shan et al. 2017). Since cities are central to net-zero emissions progress, they should and must switch to low emission development paths (Gouldson et al. 2016), which needs a sound understanding of city-level CEE dynamics across the country. Besides, studies based on city-level data, compared to those based on provincial- or sectoral-level data, deliver more detailed and robust information because they include intra-provincial differences and changes. For a country with a total area of about 9.6 million km^2 and more than 1.4 billion population, such differences might be enormous, which are likely to result in substantial heterogeneity in CEE (Tian and Zhou 2019). Hence, city-level CEE convergence analysis is in great demand.

---

[3] http://www.news.cn/english/2021-10/24/c_1310265726.htm.

This chapter addresses this issue by analysing the CEE across China based on a total factor framework. Specifically, we first use the super-efficient SBM model to evaluate city-level CEE (Tone 2001; Tran et al. 2019). The distribution dynamics model is then applied to study the club convergence of CEE. This model includes a transfer matrix consisting of differing time length of change, extending the widely used Markov model which could only analyse one-step change (Herrerias 2012). In addition, this chapter explains the club convergence of city CEE from a spatial spillover perspective. A newly compiled panel dataset consisting of 262 cities for the 2003–2016 period is used to conduct the empirical analysis. In doing so, we provide a sound understanding of spatiotemporal dynamics of China's city-level carbon emissions.

2.2 Methodology

2.2.1 Measuring Carbon Emissions Efficiency

The super-efficient SBM model can both deal with undesirable outputs redundancy and differentiate the efficient decision-making units at the frontier (Tone 2002; Tran et al. 2019). Therefore, it has been applied by an increasing number of efficiency-measuring studies (e.g., Zhang et al. 2019; Lu et al. 2020). This chapter also employs this model to measure city-level CEE. Considering n cities which use c inputs to produce e desirable outputs and u pollutants, the CEE represented by ω is measured as follows:

$$Min\omega = \frac{\frac{1}{c}\sum_{i=1}^{c}\frac{\bar{x}}{x_{ik}}}{\frac{1}{e+u}\left\{\sum_{a=1}^{e}\frac{\overline{y^d}}{y_{ak}^d} + \sum_{b=1}^{u}\frac{\overline{y^p}}{y_{bk}^p}\right\}}$$

$$s.t.\begin{cases} \bar{x} \geq \sum_{j=1,\neq k}^{n} x_{ij}\lambda_j; \ \overline{y^d} \leq \sum_{j=1,\neq k}^{n} y_{aj}^d\lambda_j; \ \overline{y^d} \geq \sum_{j=1,\neq k}^{n} y_{bj}^d\lambda_j \\ \bar{x} \geq x_k; \ \overline{y^d} \leq y_k^d, \overline{y^p} \geq y_k^p; \\ \lambda_j \geq 0; i = 1, 2, \cdots, c; \ j = 1, 2, \cdots, n, j \neq k \\ a = 1, 2, \cdots, e; \ b = 1, 2, \cdots, u \end{cases} \quad (2.1)$$

where x, y^d and y^p mean inputs, desirable outputs and pollutants vectors, respectively. k represents the analysed decision-making units.

2.2.2 Understanding Club Convergence

This chapter adopts an extended Markov approach (Tang et al. 2021) to explore the club convergence of city-level CEE. The approach extends the Markov transition probability matrix, which shows the characteristic one-step change probability, to a fixed period. Therefore, it is able to describe the dynamics of CEE.

Considering that there are w groups of cities in terms of their CEE values, the transition probability of a z-year period can be then expressed as $P_{uv}^{t,t+z} = P\{X_{t+z} = v | X_t = u\}$. This indicates the probability that a certain city, which belongs to group u in year t, becomes a part of group v after z years when evaluating its CEE. Then, the Markov transition probability P_{uv}^z for all cities in the period T is

$$P_{uv}^z = \sum_{t=t_0}^{T-z} n_{uv}^{t,t+z} \bigg/ \sum_{t=t_0}^{T-z} n_u^t \qquad (2.2)$$

where $n_{uv}^{t,t+z}$ means the number of cities, which belong to group u in year t, become a part of group v after z years when evaluating their CEE. n_u^t implies the number of cities belonging to group u in terms of CEE in year t. Finally, the Markov transition probability matrix used (Eq. (2.3)) in this chapter is formed based on Eq. (2.2). P_{uu}^z shows the probability that group-u cities still belong to the same group after z years. Larger value implies that the club convergence level in terms of CEE is higher.

$$\begin{bmatrix} \frac{n_{11}^z}{n_1^z} & \cdots & \frac{n_{1v}^z}{n_1^z} & \cdots & \frac{n_{1k}^z}{n_1^z} \\ \frac{n_{21}^z}{n_2^z} & \cdots & \frac{n_{2v}^z}{n_2^z} & \cdots & \frac{n_{2k}^z}{n_2^z} \\ \cdots & \cdots & \cdots & \cdots & \cdots \\ \frac{n_{k1}^z}{n_k^z} & \cdots & \frac{n_{kv}^z}{n_k^z} & \cdots & \frac{n_{kk}^z}{n_k^z} \end{bmatrix} = \begin{bmatrix} P_{11}^z & \cdots & P_{1v}^z & \cdots & P_{1k}^z \\ P_{21}^z & \cdots & P_{2v}^z & \cdots & P_{2k}^z \\ \cdots & \cdots & \cdots & \cdots & \cdots \\ P_{k1}^z & \cdots & P_{kv}^z & \cdots & P_{kk}^z \end{bmatrix} \qquad (2.3)$$

This chapter explains the club convergence of CEE using a spatial spillover approach (Tang et al. 2021). The global Moran index is calculated to evaluate the spatial correlation of emissions. In this analysis, positive values of the global Moran index indicate that high- and low-CEE cities show high-high and low-low spatial accumulation, respectively. Negative values imply that high- and low-CEE cities are high-low spatial accumulated. Then, the spatial Markov model is employed. By comparing the corresponding probability values of Markov and spatial Markov transition probability matrices using the chi-square test, this chapter analyses the transfer relationship of CEE between a city and its neighbouring one. More details about the used spatial spillover approach could be found in Tang et al. (2021).

2.2.3 Data

This chapter compiles a dataset consisting of panel data from 262 prefecture-level cities in China's mainland covering the 2003–2016 period. The inputs include employment of secondary industry, capital stock and annual electricity consumption. The desirable outputs and pollutants are GDP and CO_2 emissions, respectively.

Table 2.1 Descriptive statistics

Variable	Unit	Average	Standard deviation
Employment of secondary industry	10^4 people	24.0	33.3
Capital stock	10^9 Yuan	3732.3	5562.1
Annual electricity consumption	10^9 kilowatt-hours	77.6	131.6
GDP	10^9 Yuan	1237.3	1753.0
Total carbon emissions	10^4 tons	797.9	1443.3

The employment of secondary industry and GDP data are from *China Urban Statistic Yearbook* (National Bureau of Statistics of the People's Republic of China 2013–2017). Capital stock data are calculated using the perpetual inventory method (Tang et al. 2021). Following Han et al. (2018) and Fullerton and Walke (2019), annual electricity consumption is used to reflect city-level energy consumption. City-level CO_2 emissions are estimated using the accounting approach in Han et al. (2018). GDP and capital stock data are expressed by 2003 constant price. The descriptive statistics of the used variables are shown in Table 2.1.

2.3 Empirical Results and Discussion

2.3.1 Carbon Emissions Efficiency

Figure 2.1, produced by using ArcGIS 10.0 software, displays the calculated city-level CEE results in 2003 and 2016. It reveals that cities in northwestern provinces such as Ningxia, Gansu, Shaanxi and Xinjiang have relatively lower CEE than those in the coastal provinces. Socioeconomic activities in those northwestern cities have heavily relied on the use of unrenewable energy, such as coal, for decades, which might have a negative impact on CEE. In addition, since the early 2000s, many heavy-polluting enterprises have been moved from the coastal regions to the northwestern cities as a result of the implementation of the Western Development Program, one of China's regional development programs (Cheng and Zhao 2018; Wu et al. 2021). This might increase energy use and CO_2 emissions and reduce CEE.

It should be noted that some cities in northwestern provinces (e.g., Jiuquan in Gansu Province) have achieved remarkable progress in CEE improvement. Economic development and environmental protection have a complicated history in those northwestern cities, born of the competing needs of boosting local economies and protecting their natural resources. Short-term, profit-oriented plans have often won out. Since the early 2010s, however, the Chinese government has paid greater attention to environmental protection through diverse policy actions (Chang et al. 2020). Northwestern cities, which have fragile environment and suffered from heavy pollution, become the main target of those polices. Local governments are strongly

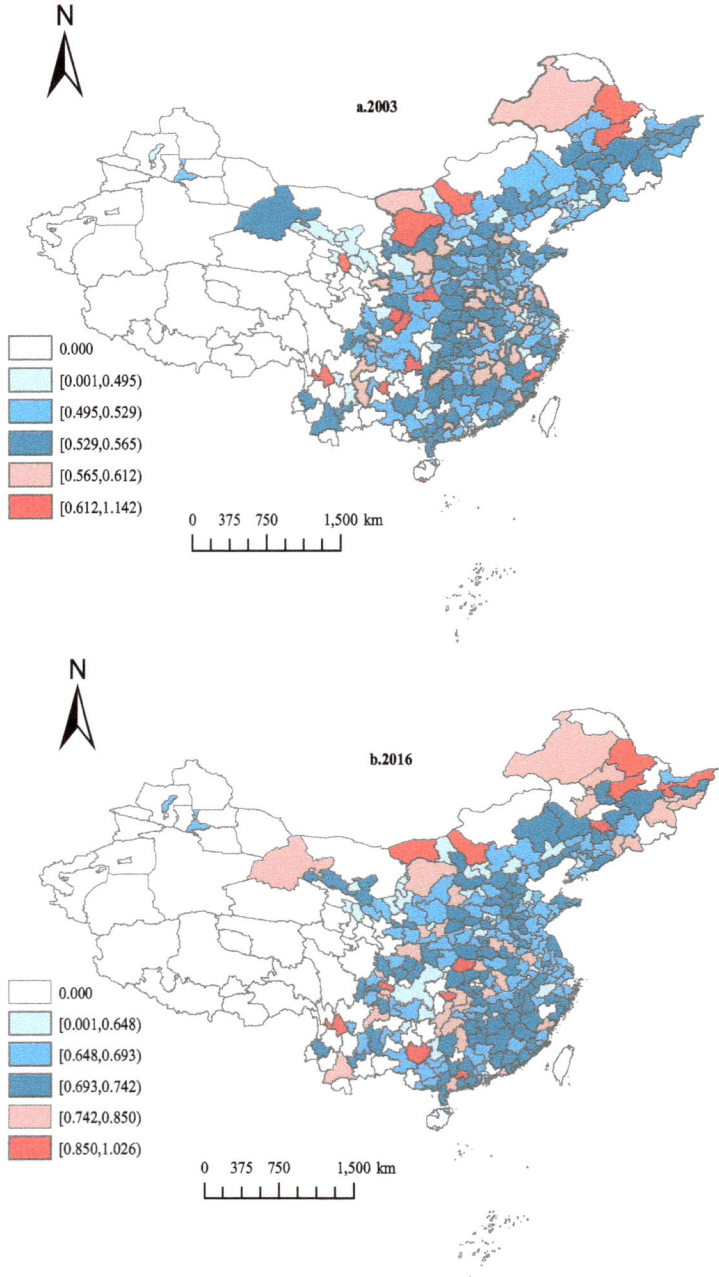

Fig. 2.1 City-level carbon emissions efficiency in China

required to take measures to adjust industrial structure, save energy and abate emissions, thus enhancing CEE.

2.3.2 Club Convergence

This chapter exams the existence of club convergence of city-level CEE within various periods. Specifically, the analysis considers five periods, ranging from one to five years in increments of one year. The cities are discretised into four groups in accordance to their CEE convergence results. According to the critical values that are 95, 100 and 105% of the mean CEE in each period, the four groups are identified as the low, intermediate-low, intermediate-high and high-level clubs. For each period, the analysis uses the Markov model to estimate the corresponding transition probability matrix of CEE. The estimated results are shown in Table 2.2.

In Table 2.2, a certain value shows the probability of the corresponding group moving from one state to another. It should be noted that the diagonal value represents the possibility that a certain group of cities will not change the state for the considered

Table 2.2 The club convergence results of city-level carbon emissions efficiency

Period	Group	Low	Intermediate-low	Intermediate-high	High
1 year	Low	0.81	0.16	0.02	0.01
	Intermediate-low	0.12	0.71	0.15	0.02
	Intermediate-high	0.01	0.17	0.71	0.11
	High	0.01	0.02	0.16	0.82
2 years	Low	0.75	0.20	0.03	0.02
	Intermediate-low	0.15	0.60	0.21	0.04
	Intermediate-high	0.03	0.23	0.59	0.16
	High	0.01	0.03	0.23	0.74
3 years	Low	0.70	0.22	0.04	0.03
	Intermediate-low	0.17	0.54	0.23	0.07
	Intermediate-high	0.05	0.25	0.50	0.20
	High	0.01	0.05	0.27	0.67
4 years	Low	0.65	0.23	0.07	0.05
	Intermediate-low	0.19	0.47	0.27	0.08
	Intermediate-high	0.09	0.27	0.41	0.24
	High	0.02	0.08	0.28	0.62
5 years	Low	0.616	0.230	0.094	0.060
	Intermediate-low	0.212	0.423	0.253	0.112
	Intermediate-high	0.086	0.294	0.364	0.256
	High	0.036	0.083	0.290	0.591

period. A diagonal value larger than 0.5 means the probability of unchanging the status is larger than that of changing, implying that the club convergence exists.

The results indicate that cities included in the low and high clubs are inclined to converge to the initial clubs within varying periods (the estimated probabilities are in the range of 0.59–0.82). This shows evidence of convergence in CEE among cities with low- and high-level CEE. On the contrary, cities with intermediate CEE are much less likely to converge to steady states since their CEE have changed rapidly. This might indicate the fierce CEE competition among them.

Here we demonstrate some potential reasons. China's low CEE cities are mostly located in the inland northwestern provinces. Most of them have fragile environmental conditions and are heavily relying on polluting industries. For example, local communities in cities such as Baotou, Tongchuan, Zhongwei and Shizuishan still need jobs and revenue provided by coal- and steel-related enterprises, thus lacking self-motivation to change. It is difficult for those cities to improve CEE substantially with current industrial structure and technologies in the short term, especially considering their relatively less developed socioeconomic conditions. Contrarily, high CEE cities, such as Heihe, Jiamusi and Lijiang, usually have more diverse or environment-base economic systems. Local communities have been enjoying benefits from the relatively sustainable industries for decades, and are likely to maintain the existing status. Similar conclusions have also been made by Liu et al. (2018) who analysed city-level industrial SO_2 and soot emissions.

This chapter also conducts the analysis with the consideration of multiple periods. Table 2.2 shows that with longer periods the probability of maintaining status becomes smaller, suggesting evidence of alleviating convergence in city-level CEE. For the five-year period, the probabilities of the low, intermediate-low, intermediate-high and high-level clubs keeping their steady states decline by 0.20, 0.29, 0.35 and 0.23, respectively. However, low- and high-level clubs still converge to their own states. The results about the intertemporal dynamics in CEE behaviour within varying periods also confirm the existence of club convergency.

2.3.3 Explaining Club Convergence from a Spatial Perspective

Further, this chapter explains the club convergency in CEE behaviour using a spatial spillover approach. The analysis calculates the global Moran index to evaluate the spatial correlation of city-level CEE (Table 2.3). The results clear indicate that high- and low-CEE cities show high-high and low-low spatial accumulation, respectively. Wang et al. (2019b) have also made similar conclusions.

We further report the Markov transfer probability matrix results for one-year and three-year periods (Tables 2.4 and 2.5). A certain value in the table represents the probability of the corresponding group moving from one state to another with given neighbouring club. The diagonal value represents the possibility that a certain group

Table 2.3 Spatial analysis results of city-level carbon emissions efficiency

Variables	Moran index value (I)	Z(I)	p-value
2003	0.07	1.91	0.06
2004	0.10	2.63	0.01**
2005	0.14	4.01	0.00***
2006	0.25	5.84	0.00***
2007	0.23	5.45	0.00***
2008	0.27	6.39	0.00***
2009	0.26	6.24	0.00***
2010	0.31	7.30	0.00***
2011	0.24	5.62	0.00***
2012	0.26	6.17	0.00***
2013	0.16	3.74	0.00***
2014	0.10	2.43	0.02*
2015	0.10	2.39	0.02*
2016	0.11	2.55	0.01*

Note: $Z(I) = \frac{1-(-1/(n-1))}{sd(I)}$, n is the sample size. The Moran index value has an asymptotic normal distribution. The critical values for Z(I) are -2.58~2.58, -1.96~1.96 and -1.65~1.65 with 1, 5 and 10% significance levels, respectively. *** and * represent 0.1% and 5% significance levels, respectively

of cities, with given neighbours, will not change the state for the considered period. By comparing the corresponding probability values of Markov and spatial Markov transition probability matrices, the transfer relationship of CEE between a city and its neighbouring one can be identified.

With low-level CEE neighbours, the probability of maintaining status for cities in the low-level group after one year (0.84) is 3.7% higher than the result ignoring spatial influence. After three years, the result (0.77) is about 10% higher than that excluding spatial effect. However, with the consideration of the influence of their low-level CEE neighbours, the possibility of transferring to intermediate-low club for those cities declines (Table 2.4). Similar conclusion can be made when a three-year period is considered. Generally, it is evident that cities with low-level CEE negatively influence their neighbours.

With high-level CEE neighbours, the probability of keeping steady status for cities in the high-level group after one year (0.84) is 2.4% higher than the result ignoring spatial influence. However, the probability is slightly smaller than the result ignoring spatial influence after three years. Overall, a city is less likely to decrease CEE if its neighbours are in the high-level club. The results after three years also indicate that cities with high-level CEE positively influence their neighbours.

By comparing the corresponding probability values of Markov and spatial Markov transition probability matrices using the chi-square test, this chapter also analyses the transfer relationship of CEE between a city and its neighbouring one and reports

Table 2.4 The spatial Markov transfer probability matrix: 1 year period

Spatial lag		1 year			
		Low	Intermediate-low	Intermediate-high	High
Low-level neighbours	Low	0.84	0.13	0.02	0.01
	Intermediate-low	0.15	0.67	0.17	0.02
	Intermediate-high	0.02	0.18	0.68	0.13
	High	0.02	0.03	0.16	0.79
Intermediate-low-level neighbours	Low	0.74	0.21	0.05	<0.01
	Intermediate-low	0.10	0.57	0.24	0.10
	Intermediate-high	<0.01	0.23	0.69	0.08
	High	<0.01	0.09	<0.01	0.91
Intermediate-high-level neighbours	Low	0.75	0.25	<0.01	<0.01
	Intermediate-low	0.16	0.77	0.07	<0.01
	Intermediate-high	0.02	0.31	0.62	0.06
	High	<0.01	<0.01	0.21	0.80
High-level neighbours	Low	0.75	0.21	0.03	0.01
	Intermediate-low	0.08	0.75	0.15	0.02
	Intermediate-high	0.01	0.15	0.74	0.10
	High	<0.01	0.01	0.16	0.84

Table 2.5 The spatial Markov transfer probability matrix: 3 years period

Spatial lag		3 years			
		Low	Intermediate-low	Intermediate-high	High
Low-level neighbours	Low	0.77	0.16	0.04	0.04
	Intermediate-low	0.19	0.47	0.27	0.07
	Intermediate-high	0.06	0.28	0.44	0.22
	High	0.03	0.08	0.25	0.65
Intermediate-low-level neighbours	Low	0.50	0.25	0.13	0.13
	Intermediate-low	0.11	0.50	0.22	0.17
	Intermediate-high	<0.01	0.27	0.55	0.18
	High	<0.01	<0.01	0.29	0.71
Intermediate-high-level neighbours	Low	0.56	0.39	0.04	0.02
	Intermediate-low	0.26	0.65	0.10	<0.01
	Intermediate-high	0.12	0.37	0.33	0.19
	High	0.03	0.03	0.37	0.57
High-level neighbours	Low	0.77	0.16	0.04	0.04
	Intermediate-low	0.19	0.47	0.27	0.07
	Intermediate-high	0.06	0.28	0.44	0.22
	High	0.03	0.08	0.25	0.65

Table 2.6 The chi-square test results for Markov and spatial Markov transition probability matrices

Duration/Year	Q value	df value	χ^2 value	P value
1	78.21	36	51.00	<0.00***
2	101.71	36	51.00	<0.00***
3	122.87	36	51.00	<0.00***
4	131.84	36	51.00	<0.00***
5	122.94	36	51.00	<0.00***

Note: *** represents 0.1% significance level

the results in Table 2.6. The results clearly show that the spatial influence in terms of CEE is significant for all periods. Moreover, Q value estimated tends to increase over time, indicating that the spatial spillover influence might be stronger.

Overall, the above results show that there is spatial correlation for Chinese cities in terms of CEE. Cities included in the low and high clubs are inclined to converge to the initial clubs within varying periods. In addition, cities with high-level CEE are likely to positively influence their neighbours, while those with low-level CEE tend to have a negative impact. Such positive or negative influence of neighbour cities could be treated as spatial spillover effect.

2.4 Conclusions

This chapter explores spatiotemporal dynamics of China's carbon emissions by conducting a city-level CEE convergence analysis. The analysis uses a newly compiled panel dataset consisting of 262 cities for the 2003–2016 period. Specifically, the city-level CEE is evaluated based on a total factor framework. The distribution dynamics model is then adopted to examine the club convergence in CEE. Contrary to previous method which could only analyse one-step change, this model includes a transfer matrix consisting of differing time length of change. Moreover, the club convergence of city CEE is explained from a spatial spillover perspective.

The empirical results reveal that cities in northwestern provinces have relatively lower CEE than those in the coastal provinces. Some northwestern cities, however, have achieved remarkable progress in CEE improvement. Besides, the results about the intertemporal dynamics in CEE behaviour within varying periods confirm the existence of club convergency. It shows evidence of convergence in CEE among cities with low- and high-level CEE. Contrarily, cities with intermediate CEE are less likely to converge to steady states. Moreover, cities with high-level CEE are likely to positively influence their neighbours, while those with low-level CEE tend to have a negative impact.

References

Barro, R. J., & Sala-i-Martin, X. (1992). Convergence. *Journal of Political Economy*, 100(2), 223-251.

Barro, R. J., Sala-i-Martin, X., Blanchard, O. J., & Hall, R. E. (1991). Convergence across states and regions. *Brookings Papers on Economic Activity*, 107.

Chang, M. C. (2020). A study on emissions efficiency, emissions technology gap ratio, room for improvement in emissions intensity, and pluralized relationships. *Environmental Science and Pollution Research*, 27(13), 14492-14502.

Cheng, A., & Zhao, F. (2018). Quantitative measure on inter-regional industry transfer and pollution transfer based on the idea of shift share analysis. *Chinese Population, Resources and Environment*, 28(5), 49-57.

Fullerton Jr, T. M., & Walke, A. G. (2019). Empirical evidence regarding electricity consumption and urban economic growth. *Applied Economics*, 51(18), 1977-1988.

Gouldson, A., Colenbrander, S., Sudmant, A., Papargyropoulou, E., Kerr, N., McAnulla, F., & Hall, S. (2016). Cities and climate change mitigation: Economic opportunities and governance challenges in Asia. *Cities*, 54, 11-19.

Haider, S., & Akram, V. (2019). Club convergence of per capita carbon emission: Global insight from disaggregated level data. *Environmental Science and Pollution Research*, 26(11), 11074-11086.

Han, F., Xie, R., Fang, J., & Liu, Y. (2018). The effects of urban agglomeration economies on carbon emissions: Evidence from Chinese cities. *Journal of Cleaner Production*, 172, 1096-1110.

Herrerias, M. J. (2012). World energy intensity convergence revisited: A weighted distribution dynamics approach. *Energy Policy*, 49, 383-399.

Iram, R., Zhang, J., Erdogan, S., Abbas, Q., & Mohsin, M. (2020). Economics of energy and environmental efficiency: Evidence from OECD countries. *Environmental Science and Pollution Research*, 27(4), 3858-3870.

Lin, B., & Wang, M. (2019). Dynamic analysis of carbon dioxide emissions in China's petroleum refining and coking industry. *Science of The Total Environment*, 671, 937-947.

Liu, C., Hong, T., Li, H., & Wang, L. (2018). From club convergence of per capita industrial pollutant emissions to industrial transfer effects: An empirical study across 285 cities in China. *Energy Policy*, 121, 300-313.

Liu, X., Zhou, D., Zhou, P., & Wang, Q. (2017). Dynamic carbon emission performance of Chinese airlines: A global Malmquist index analysis. *Journal of Air Transport Management*, 65, 99-109.

Lu, X., Chen, D., Kuang, B., Zhang, C., & Cheng, C. (2020). Is high-tech zone a policy trap or a growth drive? Insights from the perspective of urban land use efficiency. *Land Use Policy*, 95, 104583.

Mirza, F. M., & Kanwal, A. (2017). Energy consumption, carbon emissions and economic growth in Pakistan: Dynamic causality analysis. *Renewable and Sustainable Energy Reviews*, 72, 1233-1240.

National Bureau of Statistics of the People's Republic of China (2013–2017). *China Urban Statistical Yearbook*. China Statistics Press, Beijing.

Quah, D. T. (1997). Empirics for growth and distribution: stratification, polarization, and convergence clubs. *Journal of Economic Growth*, 2(1), 27-59.

Shan, Y., Guan, D., Liu, J., Mi, Z., Liu, Z., Liu, J., Schroeder, H., Cai, B., Chen, Y., Shao, S. & Zhang, Q. (2017). Methodology and applications of city level CO_2 emission accounts in China. *Journal of Cleaner Production*, 161, 1215-1225.

Tang, K., & Hailu, A. (2020). Smallholder farms' adaptation to the impacts of climate change: Evidence from China's Loess Plateau. *Land Use Policy*, 91, 104353.

Tang, K., Hailu, A., Kragt, M. E., & Ma, C. (2016). Marginal abatement costs of greenhouse gas emissions: Broadacre farming in the Great Southern Region of Western Australia. *Australian Journal of Agricultural and Resource Economics*, 60(3), 459-475.

Tang, K., He, C., Ma, C., & Wang, D. (2019). Does carbon farming provide a cost-effective option to mitigate GHG emissions? Evidence from China. *Australian Journal of Agricultural and Resource Economics*, 63(3), 575-592.

Tang, K., Xiong, C., Wang, Y., & Zhou, D. (2021). Carbon emissions performance trend across Chinese cities: Evidence from efficiency and convergence evaluation. *Environmental Science and Pollution Research*, 28(2), 1533-1544.

Tian, Y., & Zhou, W. (2019). How do CO_2 emissions and efficiencies vary in Chinese cities? Spatial variation and driving factors in 2007. *Science of the Total Environment*, 675, 439-452.

Tone, K. (2001). A slacks-based measure of efficiency in data envelopment analysis. *European Journal of Operational Research*, 130(3), 498-509.

Tone, K. (2002). A slacks-based measure of super-efficiency in data envelopment analysis. *European Journal of Operational Research*, 143(1), 32-41.

Tran, T. H., Mao, Y., Nathanail, P., Siebers, P. O., & Robinson, D. (2019). Integrating slacks-based measure of efficiency and super-efficiency in data envelopment analysis. *Omega*, 85, 156-165.

Wang, M., Wang, W., Du, S., Li, C., & He, Z. (2019a). Causal relationships between carbon dioxide emissions and economic factors: Evidence from China. *Sustainable Development*, 28(1), 73-82.

Wang, S., Shi, C., Fang, C., & Feng, K. (2019b). Examining the spatial variations of determinants of energy-related CO_2 emissions in China at the city level using Geographically Weighted Regression Model. *Applied Energy*, 235, 95-105.

Wu, J., Xu, H., & Tang, K. (2021). Industrial agglomeration, CO_2 emissions and regional development programs: A decomposition analysis based on 286 Chinese cities. *Energy*, 225, 120239.

Yang, L., Tang, K., Wang, Z., An, H., & Fang, W. (2017). Regional eco-efficiency and pollutants' marginal abatement costs in China: A parametric approach. *Journal of Cleaner Production*, 167, 619-629.

Yang, L., Yang, Y., Zhang, X., & Tang, K. (2018). Whether China's industrial sectors make efforts to reduce CO_2 emissions from production? A decomposed decoupling analysis. *Energy*, 160, 796-809.

Zhang, Y., Shen, L., Shuai, C., Tan, Y., Ren, Y., & Wu, Y. (2019). Is the low-carbon economy efficient in terms of sustainable development? A global perspective. *Sustainable Development*, 27(1), 130-152.

Zhou, D., Liang, X., Zhou, Y., & Tang, K. (2020). Does emission trading boost carbon productivity? Evidence from China's pilot emission trading scheme. *International Journal of Environmental Research and Public Health*, 17(15), 5522.

Chapter 3
Spatiotemporal Dynamics of China's Carbon Emissions: Evidence from Industrial and Regional Decoupling

Lin Yang and Kai Tang

3.1 Introduction

Since the early 1980s, China has an enviable economic growth record. Meanwhile, the country's economy is extremely relied on the consumption of fossil fuels, such as coal. During the past several decades, coal has been keeping its dominating role in China's energy mix, and its use is increasing (Fig. 3.1), thus producing enormous amounts of CO_2 emissions. Every year, China emits more CO_2 than the entire developed economies combined, and the majority of the emissions are from the industrial use of fossil fuels, mainly coal (Department of Energy Statistics 2021; Larsen et al. 2021).

Under both international and domestic pressures to address global warming, China's leaders have recently pledged to peak the country's carbon emissions before 2030, and become carbon neutral before 2060. Specifically, more 2030 pledges have been made, including increasing the share of non-fossil fuels in primary energy consumption mix to 25%, increasing the combined capacity of solar and winder power generators to 1.2 billion kilowatts, and enhancing the forest stock volume by 6 billion cubic meters from the 2005 level. However, the ambitious goal is a real challenge.

A major issue is to find a balance between moving toward carbon neutrality and stimulating economic growth (Tang and Hailu 2020; Tang and Ma 2022). China, as the world's largest developing economy, is still on its way to industrialisation and urbanisation (Tang et al. 2019; Yuan et al. 2021, 2022), and the country is expected to consume massive amounts of fossil fuels to power the process (Chen

L. Yang
School of Economics and Management, Inner Mongolia University, Huhhot 010021, China

K. Tang (✉)
School of Economics and Trade, Guangdong University of Foreign Studies, Guangzhou 510006, China
e-mail: francistang1988@hotmail.com

© The Author(s), under exclusive license to Springer Nature Singapore Pte Ltd. 2023
K. Tang (ed.), *Carbon-Neutral Pathways for China: Economic Issues*,
https://doi.org/10.1007/978-981-19-5562-4_3

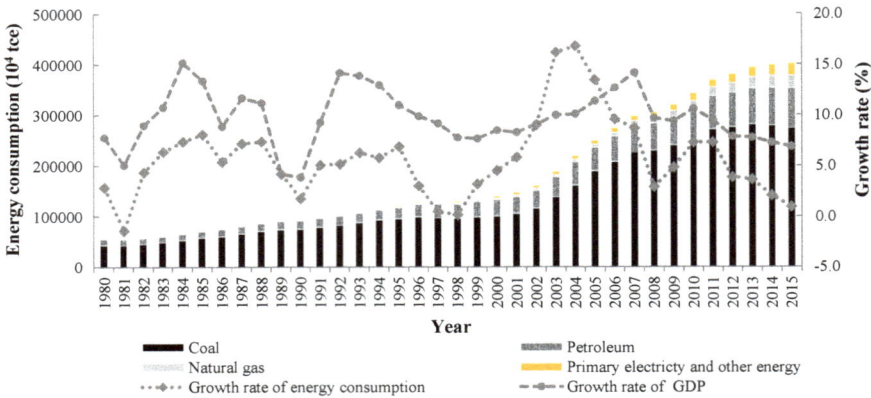

Fig. 3.1 China's energy consumption and economic growth. *Data source* China Energy Statistical Yearbook (2017)

et al. 2022; Tang et al. 2021a, b). Moreover, substantial regional and industrial heterogeneities further exacerbate the challenge. For instance, it is very difficult for some relatively less-developed regions to wean themselves from dependence on fossil-fuels-intensive industries which provide large amounts of jobs and revenue (Wu et al. 2021b). As a response, the country's policymakers have announced and are considering various differentiated policies to promote low-carbon transformation. Nevertheless, the success of these policies requires a clear understanding of spatiotemporal dynamics of the country's carbon emissions (Shahiduzzaman and Layton 2015; Tang et al. 2016; Wu et al. 2021a).

This chapter attempts to address this topic by conducting a decoupling analysis. In terms of the methods used for analysing the nexus between economic factor and carbon emissions, factor, coordination and Kuznets-Curve-related approaches have been widely applied by many recent studies in diverse contexts (e.g., Kopidou and Diakoulaki 2017; Mirza and Kanwal 2017; Zaman and Abd-el Moemen 2017; Ridzuan et al. 2020; Iqbal et al. 2021; Wu et al. 2021a). However, a common drawback of those approaches is that they could only show the overall tendency for the whole selected period but not be able to reflect the dynamics for various stages within the period. Another commonly used approach is the frontier analysis (e.g., Robaina-Alves et al. 2015; Wang and Feng 2021), which overcomes the abovementioned shortcoming well. Nevertheless, the frontier estimated is sensitive to the combination of the factors included, which might have a negative impact on the accuracy of the estimated results. The decoupling approach, by contrast, is able to identify the dynamic nexus and does not need to estimate the underlying frontier, thus providing more robust and accurate outcomes.

Originally introduced by OECD, decoupling analysis addresses the unsynchronised moving status about the economic-environmental dualistic relation. In recent years, several variants of decoupling analysis have been proposed, depicting the economic-environmental dualistic relation from differing perspectives in diverse contexts (Yang et al. 2018; Engo 2019; Xu et al. 2021). Among them, Tapio decoupling method has its own advantages (Tapio 2005); it measures entire and proportional volume changes; and it depicts the dualistic relation within an arbitrary period employing an elasticity index method, improving the flexibility and robustness of decoupling analysis (Yang et al. 2018; Song et al. 2020). Considering those advantages, some recent empirical studies have applied Tapio decoupling method at national- (Shuai et al. 2019; Li and Jiang 2020; Wang et al. 2020), regional- (Song et al. 2020; Du et al. 2021) and sectoral-level (Luo et al. 2017; Yang et al. 2018).

Nonetheless, Tapio decoupling method could not identify the drivers of environmental factors and the economic-environmental dualistic relation. A solution to this issue is to integrate it with other methods, i.e., the log-mean Divisia index (LMDI) decomposition analysis (Song et al. 2020; Yu et al. 2020; Pan et al. 2022). The combination of Tapio decoupling and LMDI decomposition is useful to explore the underlying mechanism of why the decoupling occurs in a given period (Yang et al. 2018). The recent application examples of this integrated approach in the Chinese context can be found in Zhang and Da (2015), Zhao et al. (2016), Zhou et al. (2017) and Jiang et al. (2021). Accordingly, it can be used to understand the economic-carbon dualistic relation in China from diverse perspectives.

Though some have explored the economic-carbon dynamics in China at various levels (Zhao et al. 2017; Zhang et al. 2019; Song et al. 2020; Du et al. 2021; Huang et al. 2021; Xu et al. 2021; Li et al. 2022), few has conducted a long-term systematic analysis with the consideration of both regional and industrial influencing factors. This chapter attempts to fill this gap by analysing the spatiotemporal dynamics of China's carbon emissions from industrial and regional decoupling perspectives. A decomposed decoupling approach consisting of Tapio decoupling and LMDI decomposition methods is applied to evaluate data from key industries in different regions for the 1996–2015 period. The results not only contribute to the related literature but also provide empirical evidence for refining and designing regional and industrial carbon policies for the global top carbon emitter.

3.2 Methodology

3.2.1 The Log-Mean Divisia Index (LMDI) Decomposition Method

LMDI decomposition method, one of index decomposition approaches, is adopted in this chapter to identify the drivers of carbon emissions and the economic-carbon dualistic relation. Compared with structural decomposition methods and other index decomposition approaches, LMDI decomposition method has the advantage of path independency, aggregation consistency and complete decomposition (Ang 2005,

2015). As an improved variant of Kaya identity (Kaya 1990), LMDI decomposition method addresses the connections among carbon emissions, economic output, energy use and population as follows.

$$
\begin{aligned}
Carbon &= \sum_i \sum_j Carbon_{ij} \\
&= \sum_i \sum_j \frac{Carbon_{ij}}{Energy_{ij}} \cdot \frac{Energy_{ij}}{Energy_i} \cdot \frac{Energy_i}{GDP_i} \cdot \frac{GDP_i}{GDP} \cdot \frac{GDP}{Population} \cdot Population \\
&= \sum_i \sum_j CC_{ij} \cdot ES_{ij} \cdot EI_i \cdot GS_i \cdot GDPP \cdot Population \quad (3.1)
\end{aligned}
$$

In Eq. (3.1), $Carbon$ denotes the total CO_2 emissions in a region; $Carbon_{ij}$ represents the CO_2 emissions caused by the use of energy j in industry i; $Energy_{ij}$ represents the use of energy j in industry i; $Energy_i$ denotes the total energy use of industry i; GDP_i means the economic output of industry i; GDP is the overall economic output in the considered region; $Population$ shows the regional population; CC_{ij} denotes CO_2 emissions coefficient of energy j in industry i; ES_{ij} represents the energy mix of industry i; EI_i represents the energy intensity of industry i; GS_i denotes the share of the economic output of industry i in the overall economic output of the considered region, implying the regional industrial structure; $GDPP$ means the per capita economic output.

The CO_2 emissions change in a region between period 0 and period M is then expressed as:

$$
\begin{aligned}
\Delta Carbon &= \sum_i \sum_j CC_{ij}^M \cdot ES_i^M \cdot EI_i^M \cdot GS_i^M \cdot GDPP^M \cdot Population^M \\
&\quad - \sum_i \sum_j CC_{ij}^0 \cdot ES_i^0 \cdot EI_i^0 \cdot GS_i^0 \cdot GDPP^0 \cdot Population^0 \quad (3.2)
\end{aligned}
$$

The overall effects of all variables on changes of total CO_2 emissions could be decomposed into the sum of effects of each potential determinant. Specifically, the effects include emission coefficient, energy structure, energy intensity, industrial structure, per capita economic output, and population effects. For example, emission coefficient effect (ΔCC) can be expressed as:

$$
\Delta CC = \sum_i \sum_j L\left(Carbon_{ij}^M, Carbon_{ij}^0\right) \ln\left(\frac{CC_{ij}^M}{CC_{ij}^0}\right)
$$

$$
\begin{aligned}
&L\left(Carbon_{ij}^M, Carbon_{ij}^0\right) \\
&= \begin{cases} (Carbon_{ij}^M - Carbon_{ij}^0)/(\ln Carbon_{ij}^M - \ln Carbon_{ij}^0), Carbon_{ij}^M \neq Carbon_{ij}^0 \\ Carbon_{ij}^M, Carbon_{ij}^M = Carbon_{ij}^0 \\ 0, Carbon_{ij}^M = Carbon_{ij}^0 = 0 \end{cases}
\end{aligned} \quad (3.3)
$$

More details about other effects and the LMDI decomposition method can be found in Ang (2015). It should be noted that the emission coefficients used are from IPCC (2006) and have constant values, implying that $\Delta CC = 0$. Therefore, Eq. (3.2) is then rewritten as follows:

$$\Delta Carbon = \Delta ES + \Delta EI + \Delta GS + \Delta GDPP + \Delta Population \qquad (3.4)$$

3.2.2 Tapio Decoupling Method

This chapter follows Tapio (2005) to conduct a Tapio decoupling analysis. In a region, the decoupling elasticity index of CO_2 emissions from economy (e) is:

$$e = \frac{\%\Delta C}{\%\Delta GDP} = \frac{(Carbon_M - Carbon_0)/Carbon_0}{(GDP_M - GDP_0)/GDP_0} = \frac{\Delta Carbon/Carbon}{\Delta GDP/GDP} \qquad (3.5)$$

where $\Delta Carbon$ and ΔGDP denote CO_2 emissions and GDP increments, respectively. Following Tapio (2005) and Pan et al. (2019), the degrees of coupling and decoupling can be divided into eight areas, as shown in Fig. 3.2.

Considering Eqs. (3.4) and (3.5) can be extended as:

$$e = \frac{\Delta Carbon/Carbon}{\Delta GDP/GDP} = \frac{(\Delta ES + \Delta EI + \Delta GS + \Delta GDPP + \Delta Population)/Carbon}{\Delta GDP/GDP}$$

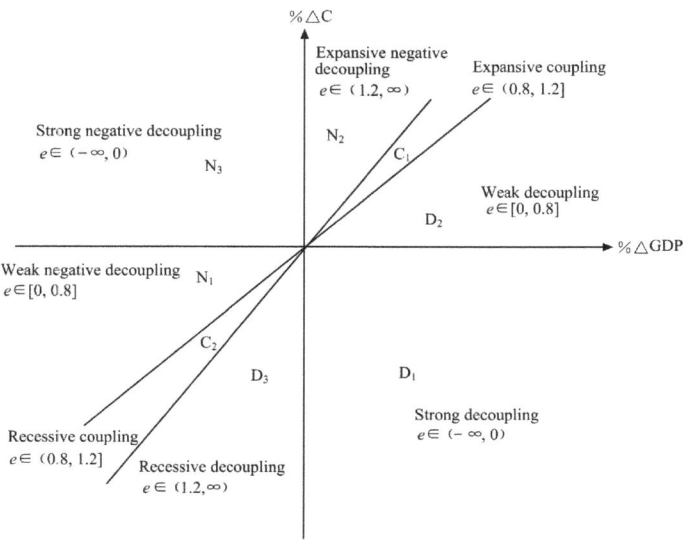

Fig. 3.2 The degrees of coupling and decoupling of CO_2 emissions from economy. *Figure source* Pan et al. (2019)

$$= \frac{\Delta ES}{Carbon} \times \frac{GDP}{\Delta GDP} + \frac{\Delta EI}{Carbon} \times \frac{GDP}{\Delta GDP} + \frac{\Delta GS}{Carbon} \times \frac{GDP}{\Delta GDP}$$
$$+ \frac{\Delta GDPP}{Carbon} \times \frac{GDP}{\Delta GDP} + \frac{\Delta Population}{Carbon} \times \frac{GDP}{\Delta GDP}$$
$$= e_{ES} + e_{EI} + e_{GS} + e_{GDPP} + e_{Population} \tag{3.6}$$

where e_{ES}, e_{EI}, e_{GS}, e_{GDPP} and $e_{Population}$ denote energy structure, energy intensity, industrial structure, per capita economic output, and population elasticity indexes, respectively.

3.2.3 Data

This chapter compiles a dataset consisting of panel data from 30 provincial-level regions in China covering the 1996–2015 period. Due to data availability, Tibet, Hong Kong, Macau and Taiwan are not considered in the analysis. Five main industries, including the manufacturing, commercial, construction, transportation and agriculture, are investigated. The analysis uses industrial added value data, expressed by 1996 constant price, to show the industrial economic output of industry. Industrial added value, industrial and population data are from *China Statistical Yearbook* (1997–2016). Energy data are from *China Energy Statistical Yearbook* (1997–2016). The analysis considers industrial energy-related CO_2 emissions, which represent the majority of the country's emissions (Wu et al. 2021b). The used calculation methods and coefficients are from IPCC (2006).

3.3 Results and Discussion

3.3.1 China's Industrial CO_2 Emissions

During the 1996–2015 period, China's total industrial energy-related CO_2 emissions had increased more than 2.55 times, reaching approximately 8.5 billion tonnes in 2015. Emissions from all studied industries boosted. The manufacturing industry was the dominating emitter, accounting for around 80% of the total emissions. The second largest emitter was the transportation industry, and its emissions had increased more than 5 times over two decades. The CO_2 emissions from construction industry accounted for the smallest share (less than 2%) during the whole study period.

Overall, striking regional heterogeneity existed among provinces in terms of industrial CO_2 emissions. Top emitters were five coaster provinces including Hebei, Shandong, Jiangsu, Liaoning and Guangdong (Fig. 3.3). Specifically, Guangdong's GDP and Jiangsu's GDP rank the first and the second in China, respectively. Those two provinces are also among the most populous provinces. The economies of

Hebei, Shandong and Liaoning rely heavily on energy-intensive industries, such as iron and steel, non-ferrous metals and petrochemical industries, thus generating CO_2 enormous emissions. On the contrary, provinces with the smallest emissions include Qinghai and Hainan. Both of them are less densely populated, and the local economic activities are based largely on eco-related industries, thus emitting relatively less emissions. All provinces, expect Beijing, experienced increased emissions. The declining emissions in Beijing should be the result of the restructured local economy; emissions-intensive enterprises, which are forced to curb or even halt operations, are replaced by greener enterprises in order to improve air quality and meet the carbon reduction target.

Despite their relatively smallish amounts of CO_2 emissions, provinces like Ningxia, Qinghai and Hainan had rapid emissions growth. Specifically, Ningxia's industrial energy-related CO_2 emissions had increased more than 7 times over two decades, and the share increased from 0.4% to 1.3%. Such a change is partly attributed to the local growth needs and the national industrial transferring programs that inland provinces like Ningxia and Qinghai undertake the primary responsibility for energy production and supply. Wu et al. (2021b) argued that those transferring programs might have decreased the environmental efficiency gains from industrial agglomeration during the country's ongoing urbanisation and industrialisation process.

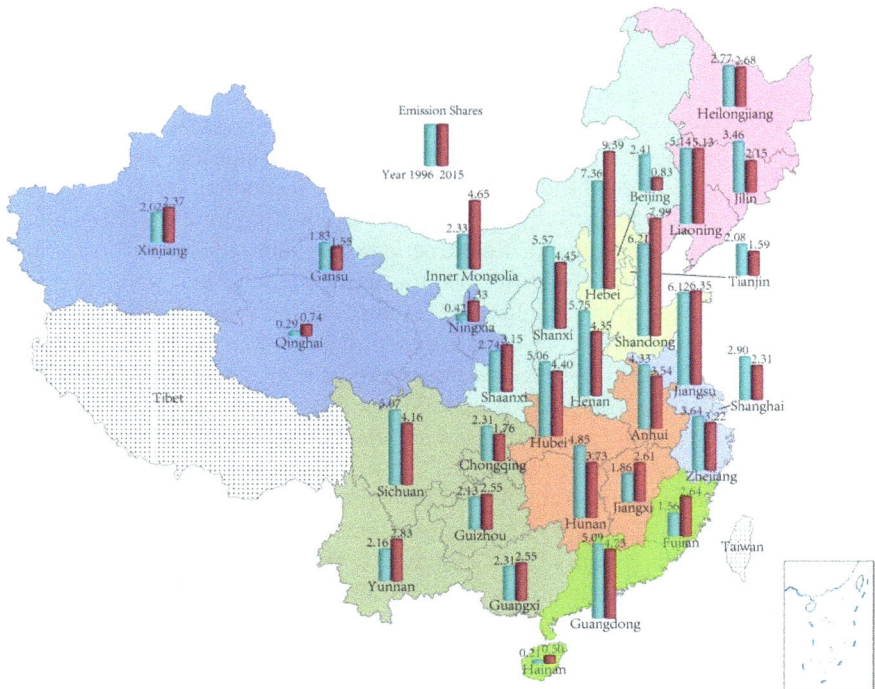

Fig. 3.3 CO_2 emissions shares by province (%)

Whereas, the too-rapid growth rate poses a challenge to the country's attempt to achieve net-zero emissions in coming decades.

3.3.2 The Decoupling of Industrial CO_2 Emissions

The results about the decoupling elasticity index of CO_2 emissions from economy (Eq. 3.5) are shown in Figs. 3.4 and 3.5. Considering that no economic recession occurred in China during the studied years, four out of eight degrees of coupling and decoupling might occur, including strong, weak and expansive negative decoupling states and expansive coupling state. Overall, the studied industries, especially the manufacturing industry, generally exhibited weak decoupling over the studied years. To be more specific, the decoupling status for the whole country appeared an inverted-U shape; strong decoupling in the 1996–2000 and the 2011–2015 periods, expansive coupling in the 2001–2005 period and weak decoupling in the 2006–2010 period.

The manufacturing, commercial and transportation industries also had the similar inverted-U shape in terms of the decoupling degrees. Among five industries, the manufacturing industry is the dominating source of economic output in many provinces and the main carbon emitter for the whole country (53.27% in 2017) (IPCC 2019). The result also affirms that the manufacturing industry has a determining influence on the decoupling progress of China's industries. For the manufacturing industry, its proportion for economic output increased by 8%, while that for carbon emissions decreased by more than 5% over the entire period. Strong decoupling status mainly appeared in the 2011–2015 period, indicating that CO_2 emissions from this industry had been reduced. For the construction industry, its decoupling status was characterised by the N shape. However, the decoupling states of agricultural and transportation industries displayed weak decoupling for the whole studied period, which supports the claim that more actions should be taken to reduce carbon emissions from agricultural production and transportation in China.

The estimated decoupling results also revealed strong regional heterogeneity. In general, the performance of southeastern provinces was better than that of northeastern ones. Most industries in southeastern provinces had declining decoupling elasticity indexes, while the agricultural and construction industries in northeastern provinces Heilongjiang and Jilin had relatively high decoupling elasticity values, exhibiting expansive negative decoupling. A tendency toward coupling occurred in Heilongjiang, Gansu, Qinghai and Ningxia. The manufacturing industries in Qinghai and Ningxia and commercial industry in Gansu were still characterised by high carbon emissions. Contrarily, other provinces ultimately exhibited weak or strong decoupling. Specifically, several energy- and heavy-industry-based provinces like Shandong and Xinjiang ultimately achieved strong decoupling mainly due to the low-emission transformation of their manufacturing industries, showing that those provinces had limited industrial CO_2 emissions remarkably with expanding production during the studied years (Fig. 3.4). For several coaster provinces including Liaoning, Beijing, Shandong, Shanghai, Guangxi and one inland

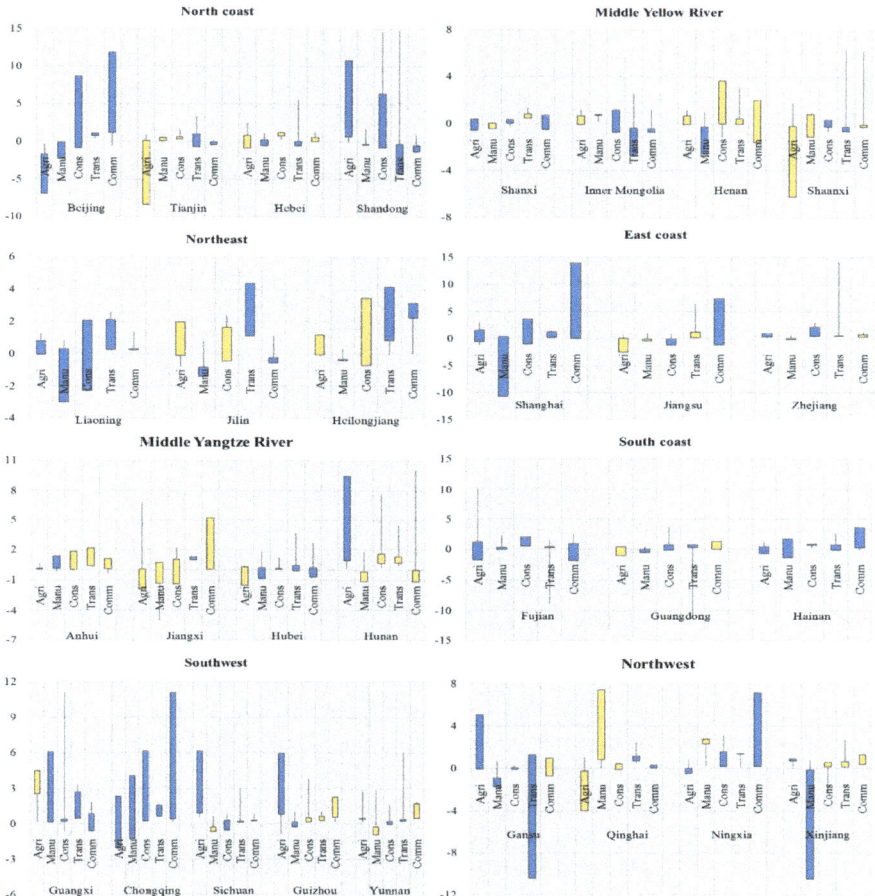

Fig. 3.4 The decoupling elasticity results of industrial CO_2 emissions from economy (1996–2015). *Note* Yellow pillar indicates that the decoupling index in 1996 was lower than that in 2015, while blue pillar means the opposite. The top and bottom of the bar represent the highest and lowest decoupling index during the whole period, respectively

province Chongqing, their decoupling elasticity indexes for all studied industries declined significantly.

3.3.3   Decomposition Results for Industrial CO_2 Emissions

The evaluated decomposition results show that economic output and population factors were the drivers for China's CO_2 industrial emissions, and economic output continued playing the dominating role over the studied years. Averagely, economic

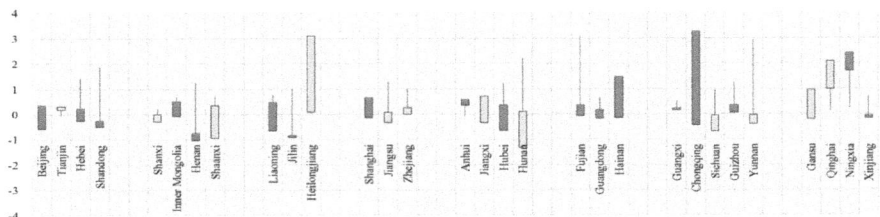

Fig. 3.5 The regional decoupling elasticity results of industrial CO_2 emissions from economy (1996–2015). *Note* Yellow pillar indicates that the decoupling index in 1996 was lower than that in 2015, while blue pillar means the opposite. The top and bottom of the bar represent the highest and lowest decoupling index during the whole period, respectively

output accounted for about 91.3% of the country's emissions during the studied years. China's economic activities have been closely accompanied by the massive consumption of fossil fuels (Fig. 3.1). During the past several decades, the country's use of fossil fuels has been increasing with the continuous expanding economic output, thus resulting in enormous amounts of CO_2 emissions. Moreover, considering that China is highly likely to meet most of its energy needs with relying on fossil fuels[1] and keep its economy growing in the short- and middle-term, economic output is expected to continue playing the key role in driving emissions. Population factor accounted for about 6.7% of the country's emissions during the studied years. Raising living standards further boosts demands for goods and services which may result in consuming more resources and generating more CO_2 emissions. However, it should be noted that such an effect might be weaker since the country's population will peak in the next few years.[2]

On the contrary, energy intensity contributed most to the emissions reduction, resulting in about 336 million tonne reduction for the whole period. In China, energy intensity is generally highly influenced by energy technologies and regulations. Since the Chinese government has placed a priority on investing in clean energy technologies, primarily because they enable the country to tackle serious air pollution problem, and launched a series of more stringent environmental regulations,[3] it is highly likely that the reduction effect of energy intensity will be greater, at least in the short- and middle-term.

Compared with other factors, the effects of energy structure and industrial structure were relatively weak. In terms of energy structure, in China, coal, one of the most polluting forms of energy, has been the dominant source of energy for decades,

---

[1] https://www.cnbc.com/2021/04/29/climate-china-has-no-other-choice-but-to-rely-on-coal-power-for-now.html.

[2] The United Nations predicts the population of mainland China will peak in 2030 before decreasing. See: https://www.reuters.com/world/china/china-2020-census-shows-slowest-population-growth-since-1-child-policy-2021-05-11/

[3] During the 2015–2020 period, China had a national target for a reduction in energy intensity of 15%. The latest Five Year Plan outline for the 2021–2025 period targets a further 13.5% cut by 2025. See http://www.news.cn/english/2021-11/09/c_1310300766.htm for more details.

and its use is still increasing (Fig. 3.1). In 2020, the share of coal in primary energy consumption mix was 56.8%, while the share was 72.4% in 2005 and roughly 70% in 2010. The dominant coal was followed by oil (20%), clean electricity (hydro, wind, solar and nuclear) (9%), natural gas (8%), and biomass (3%) (Department of Energy Statistics, 2021). Globally, China accounts for more than half of the world's total coal use.[4] The country is likely to rely on coal to support economic growth and expects the fuel to remain in the energy mix long, thus resulting in the weak effect of energy structure.

This chapter also explores the decomposition results for industrial CO_2 emissions from a regional perspective. In terms of the decoupling effort index, the evaluated results convince that China's provincial regions were trying to decouple CO_2 emissions and growth (Fig. 3.6). However, clear regional disparity existed. The country's capital, Beijing, had the highest decoupling effect index (1.25), which was followed by Jilin (1.00) and Henan (0.99). On the contrary, two northwestern provinces, Qinghai and Ningxia, whose decoupling effort index values were less than 0.1, performed worst during the studied period. For most provincial regions, the decoupling efforts index moved along a U-shaped curve. In terms of the influencing factors of emissions, the results consistently demonstrate that energy intensity contributed most to the decoupling process (Fig. 3.7). However, this process was challenged by economic output and population factors.

3.4 Conclusions

This chapter analyses the spatiotemporal dynamics of China's carbon emissions from industrial and regional decoupling perspectives. A decomposed decoupling approach consisting of Tapio decoupling and LMDI decomposition methods is applied to evaluate data from key industries in different regions for the 1996–2015 period.

The results show that China's total industrial energy-related CO_2 emissions had increased more than 2.55 times during the studied period, reaching approximately 8.5 billion tonnes in 2015. The manufacturing industry accounted for around 80% of the total emissions. All provinces, expect Beijing, experienced increased emissions. In general, the performance of southeastern provinces was better than that of northeastern ones. Most industries in southeastern provinces had declining decoupling elasticity indexes, while the agricultural and construction industries in Heilongjiang and Jilin had relatively high decoupling elasticity values. Economic output and population factors were the drivers for China's CO_2 emissions, and economic output continued playing the dominating role over the studied years. Energy intensity contributed most to the emissions reduction, resulting in about 336 million tonne reduction for the whole period. Overall, the results argue that China's provincial regions were trying to decouple CO_2 emissions and growth.

---

[4] See https://ember-climate.org/global-electricity-review-2021/g20-profiles/china/ for more information.

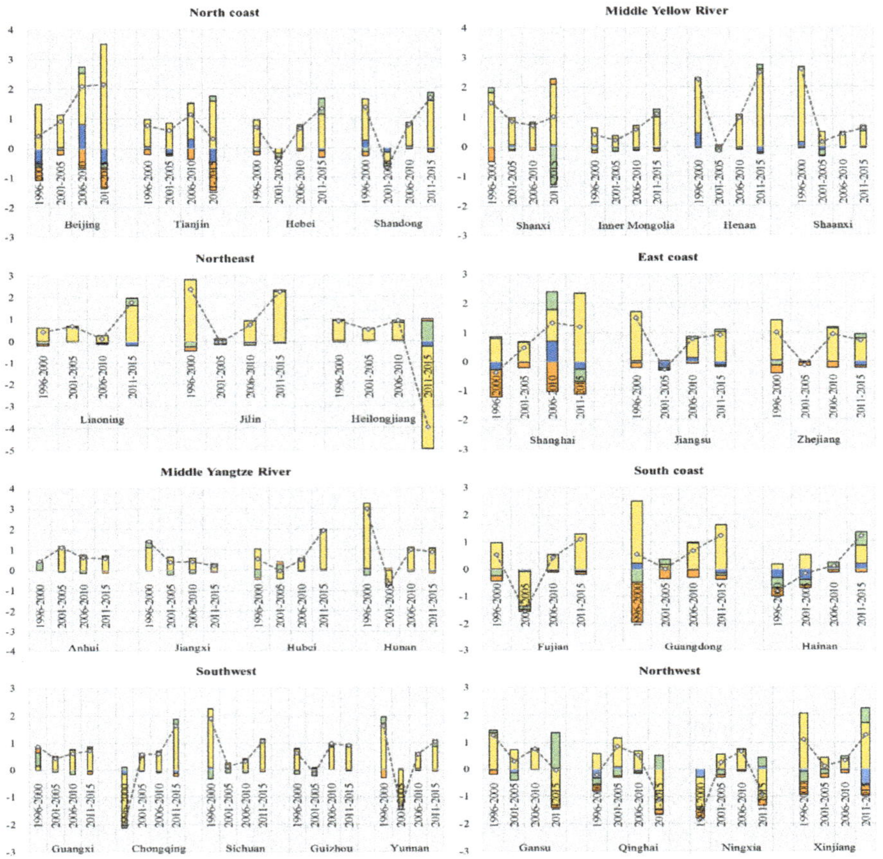

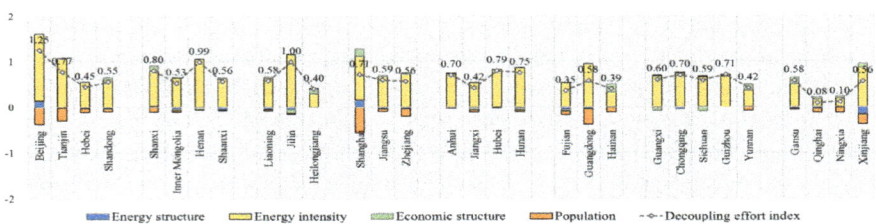

Fig. 3.6 Regional results about decoupling effort index decomposition

Fig. 3.7 Regional results about decoupling effort index decomposition, average level

References

Ang, B. W. (2005). The LMDI approach to decomposition analysis: A practical guide. *Energy Policy*, 33(7), 867-871.

Ang, B. W. (2015). LMDI decomposition approach: A guide for implementation. *Energy Policy*, 86, 233-238.

Chen, X., Chen, G., Lin, M., Tang, K., & Ye, B. (2022). How does anti-corruption affect enterprise green innovation in China's energy-intensive industries?. *Environmental Geochemistry and Health*, 44, 2919-2942.

Department of Energy Statistics, National Bureau of Statistics (2021). *China Energy Statistical Yearbook 2021*. China Statistics Press, Beijing.

Du, X., Shen, L., Wong, S. W., Meng, C., & Yang, Z. (2021). Night-time light data based decoupling relationship analysis between economic growth and carbon emission in 289 Chinese cities. *Sustainable Cities and Society*, 73, 103119.

Engo, J. (2019). Decoupling analysis of CO_2 emissions from transport sector in Cameroon. *Sustainable Cities and Society*, 51, 101732.

Huang, Y., Yu, Q., & Wang, R. (2021). Driving factors and decoupling effect of carbon footprint pressure in China: Based on net primary production. *Technological Forecasting and Social Change*, 167, 120722.

IPCC. (2006). *Greenhouse Gas Inventory: IPCC Guidelines for National Greenhouse Gas Inventories*. United Kingdom Meteorological Office, Bracknell, England.

IPCC. (2019). *Refinement to the 2006 IPCC Guidelines for National Greenhouse Gas Inventories*. https://www.ipcc-nggip.iges.or.jp/public/2019rf/index.html

Iqbal, N., Abbasi, K. R., Shinwari, R., Guangcai, W., Ahmad, M., & Tang, K. (2021). Does exports diversification and environmental innovation achieve carbon neutrality target of OECD economies?. *Journal of Environmental Management*, 291, 112648.

Jiang, J., Zhao, T., & Wang, J. (2021). Decoupling analysis and scenario prediction of agricultural CO_2 emissions: An empirical analysis of 30 provinces in China. *Journal of Cleaner Production*, 320, 128798.

Kaya, Y. (1990). Impact of carbon dioxide emission control on GNP growth: Interpretation of proposed scenarios. In: Presented at the IPCC Energy and Industry Subgroup, Response Strategies Working Group, Paris.

Kopidou, D., & Diakoulaki, D. (2017). Decomposing industrial CO_2 emissions of Southern European countries into production-and consumption-based driving factors. *Journal of Cleaner Production*, 167, 1325-1334.

Larsen, K., Pitt, H., Grant, M., & Houser, T. (2021). China's greenhouse gas emissions exceeded the developed world for the first time in 2019. Rhodium Group. https://rhg.com/research/chinas-emissions-surpass-developed-countries/#_ftn1

Li, R., & Jiang, R. (2020). Investigating effect of R&D investment on decoupling environmental pressure from economic growth in the global top six carbon dioxide emitters. *Science of The Total Environment*, 740, 140053.

Li, K., Ma, M., Xiang, X., Feng, W., Ma, Z., Cai, W., & Ma, X. (2022). Carbon reduction in commercial building operations: A provincial retrospection in China. *Applied Energy*, 306, 118098.

Luo, Y., Long, X., Wu, C., & Zhang, J. (2017). Decoupling CO_2 emissions from economic growth in agricultural sector across 30 Chinese provinces from 1997 to 2014. *Journal of Cleaner Production*, 159, 220-228.

Mirza, F. M., & Kanwal, A. (2017). Energy consumption, carbon emissions and economic growth in Pakistan: Dynamic causality analysis. *Renewable and Sustainable Energy Reviews*, 72, 1233-1240.

National Bureau of Statistics of the People's Republic of China (1997–2016). *China Statistical Yearbook*. China Statistics Press, Beijing.

National Bureau of Statistics of the People's Republic of China (1997–2016). *China Energy Statistical Yearbook*. China Statistics Press, Beijing.

Pan, W. L., Pan, W., Liu, S., Tsai, S. B., Hu, C., & Tu, H. (2019). China's provincial energy-related carbon emissions-economy nexus: A two-stage framework based on decoupling analysis and panel vector autoregression. *Energy Science & Engineering*, 7(4), 1201-1213.

Pan, X., Guo, S., Xu, H., Tian, M., Pan, X., & Chu, J. (2022). China's carbon intensity factor decomposition and carbon emission decoupling analysis. *Energy*, 239, 122175.

Ridzuan, N. H. A. M., Marwan, N. F., Khalid, N., Ali, M. H., & Tseng, M. L. (2020). Effects of agriculture, renewable energy, and economic growth on carbon dioxide emissions: Evidence of the environmental Kuznets curve. *Resources, Conservation and Recycling*, 160, 104879.

Robaina-Alves, M., Moutinho, V., & Macedo, P. (2015). A new frontier approach to model the eco-efficiency in European countries. *Journal of Cleaner Production*, 103, 562-573.

Shahiduzzaman, M., & Layton, A. (2015). Changes in CO_2 emissions over business cycle recessions and expansions in the United States: A decomposition analysis. *Applied Energy*, 150, 25-35.

Shuai, C., Chen, X., Wu, Y., Zhang, Y., & Tan, Y. (2019). A three-step strategy for decoupling economic growth from carbon emission: Empirical evidences from 133 countries. *Science of the total environment*, 646, 524-543.

Song, Y., Sun, J., Zhang, M., & Su, B. (2020). Using the Tapio-Z decoupling model to evaluate the decoupling status of China's CO_2 emissions at provincial level and its dynamic trend. *Structural Change and Economic Dynamics*, 52, 120-129.

Tang, K., & Hailu, A. (2020). Smallholder farms' adaptation to the impacts of climate change: Evidence from China's Loess Plateau. *Land Use Policy*, 91, 104353.

Tang, K., & Ma, C. (2022). The cost-effectiveness of agricultural greenhouse gas reduction under diverse carbon policies in China. *China Agricultural Economic Review*. https://doi.org/10.1108/CAER-01-2022-0008

Tang, K., He, C., Ma, C., & Wang, D. (2019). Does carbon farming provide a cost-effective option to mitigate GHG emissions? Evidence from China. *Australian Journal of Agricultural and Resource Economics*, 63(3), 575-592.

Tang, K., Liu, Y., Zhou, D., & Qiu, Y. (2021a). Urban carbon emission intensity under emission trading system in a developing economy: Evidence from 273 Chinese cities. *Environmental Science and Pollution Research*, 28(5), 5168-5179.

Tang, K., Xiong, C., Wang, Y., & Zhou, D. (2021b). Carbon emissions performance trend across Chinese cities: Evidence from efficiency and convergence evaluation. *Environmental Science and Pollution Research*, 28(2), 1533-1544.

Tang, K., Zhou, Y., Liang, X., & Zhou, D. (2021c). The effectiveness and heterogeneity of carbon emissions trading scheme in China. *Environmental Science and Pollution Research*, 28(14), 17306-17318.

Tang, K., Yang, L., & Zhang, J. (2016). Estimating the regional total factor efficiency and pollutants' marginal abatement costs in China: A parametric approach. *Applied Energy*, 184, 230-240.

Tapio, P. (2005). Towards a theory of decoupling: Degrees of decoupling in the EU and the case of road traffic in Finland between 1970 and 2001. *Transport Policy*, 12(2), 137-151.

Wang, C., Wood, J., Geng, X., Wang, Y., Qiao, C., & Long, X. (2020). Transportation CO_2 emission decoupling: Empirical evidence from countries along the belt and road. *Journal of Cleaner Production*, 263, 121450.

Wang, M., & Feng, C. (2021). The consequences of industrial restructuring, regional balanced development, and market-oriented reform for China's carbon dioxide emissions: A multi-tier meta-frontier DEA-based decomposition analysis. *Technological Forecasting and Social Change*, 164, 120507.

Wu, J., Feng, Z., & Tang, K. (2021a). The dynamics and drivers of environmental performance in Chinese cities: A decomposition analysis. *Environmental Science and Pollution Research*, 28, 30626–30641.

Wu, J., Xu, H., & Tang, K. (2021b). Industrial agglomeration, CO_2 emissions and regional development programs: A decomposition analysis based on 286 Chinese cities. *Energy, 225*, 120239.

Xu, W., Xie, Y., Xia, D., Ji, L., & Huang, G. (2021). A multi-sectoral decomposition and decoupling analysis of carbon emissions in Guangdong province, China. *Journal of Environmental Management, 298*, 113485

Yang, L., Yang, Y., Zhang, X., & Tang, K. (2018). Whether China's industrial sectors make efforts to reduce CO_2 emissions from production? A decomposed decoupling analysis. *Energy, 160*, 796-809.

Yu, J., Shao, C., Xue, C., & Hu, H. (2020). China's aircraft-related CO_2 emissions: Decomposition analysis, decoupling status, and future trends. *Energy Policy, 138*, 111215.

Yuan, F., Tang, K., & Shi, Q. (2021). Does Internet use reduce chemical fertilizer use? Evidence from rural households in China. *Environmental Science and Pollution Research, 28*(5), 6005-6017.

Yuan, F., Tang, K., Shi, Q., Qiu, W., & Wang, M. (2022). Rural women and chemical fertiliser use in rural China. *Journal of Cleaner Production, 344*, 130959.

Zaman, K., & Abd-el Moemen, M. (2017). Energy consumption, carbon dioxide emissions and economic development: Evaluating alternative and plausible environmental hypothesis for sustainable growth. *Renewable and Sustainable Energy Reviews, 74*, 1119-1130.

Zhang, X., Zhao, Y., Wang, C., Wang, F., & Qiu, F. (2019). Decoupling effect and sectoral attribution analysis of industrial energy-related carbon emissions in Xinjiang, China. *Ecological Indicators, 97*, 1-9.

Zhang, Y. J., & Da, Y. B. (2015). The decomposition of energy-related carbon emission and its decoupling with economic growth in China. *Renewable & Sustainable Energy Reviews, 41*, 1255-1266.

Zhao, X., Zhang, X., Li, N., Shao, S., & Geng, Y. (2017). Decoupling economic growth from carbon dioxide emissions in China: A sectoral factor decomposition analysis. *Journal of Cleaner Production, 142*, 3500-3516.

Zhao, X., Zhang, X., & Shao, S. (2016). Decoupling CO_2, emissions and industrial growth in China over 1993–2013: The role of investment. *Energy Economics, 60*, 275-292.

Zhou, X., Zhang, M., Zhou, M., & Zhou, M. (2017). A comparative study on decoupling relationship and influence factors between China's regional economic development and industrial energy–related carbon emissions. *Journal of Cleaner Production, 142*, 783-800.

Chapter 4
Provincial Carbon Reduction Costs and Potentials in China: A Total Factor Analysis

Kai Tang and Lin Yang

4.1 Introduction

Global economic activities still highly rely on the use of energy and natural resources in contemporary society. Such a nexus is inevitably accompanied by massive emissions of pollutants, i.e., CO_2 and SO_2 (Kaufmann et al. 1998; Canadell et al. 2007; Antonakakis et al. 2017; Ai et al. 2021). In developing economies, partly due to the continuous demand for boosting economy and improving living standards and relatively undeveloped production technologies, those related emissions of pollutants might be even more enormous (Kaufmann et al. 1998; Sarkodie and Strezov 2018; Yuan et al. 2021, 2022; Chen et al. 2022; Liu et al. 2022). For example, China, the world's largest developing economy, has been the largest emitter of GHG in the world since 2006. Every year, China emits more GHG than the entire developed countries combined. In 2019, China emitted 27% of the world's GHG, while the US and India contributed 11% and 6%, respectively.

Under both international and domestic pressures to do more to address global warming, China has taken measures to combat climate change. Recently, a serious of 2030 pledges have been made, including decreasing carbon intensity, the amount of CO_2 emissions per unit of GDP, by 65% from the 2005 level. Multiple strategies, such as enhancing energy efficiency, optimizing energy mix and adjusting industrial structure, have also been adopted to slow the growth of the country's GHG emissions (Tang et al. 2021c). Nevertheless, those measures, which aim at achieving a better balance between ensuring economic growth and combating climate change, are bound and costly for the whole society (Tang et al. 2018, 2019; Tang and Ma 2022).

K. Tang (✉)
School of Economics and Trade, Guangdong University of Foreign Studies, Guangzhou 510006, China
e-mail: francistang1988@hotmail.com

L. Yang
School of Economics and Management, Inner Mongolia University, Huhhot 010021, China

© The Author(s), under exclusive license to Springer Nature Singapore Pte Ltd. 2023   49
K. Tang (ed.), *Carbon-Neutral Pathways for China: Economic Issues*,
https://doi.org/10.1007/978-981-19-5562-4_4

Both policymakers and researchers generally agree that without radical technological changes in production systems, a trade-off between expanding output and reducing GHG emissions is unavoidable and the potential of reduction is constrained considering that the available resources and technologies are limited (Tang et al. 2016a; Yang et al. 2017; Doelman et al. 2020; Rosenzweig et al. 2020). Therefore, it is necessary to have a sound understanding of the costs and potentials associated with the reduction efforts, contributing to designing and refining economic and climate policy regimes.

Emerging studies have argued that the distance function approach might be useful to derive the reduction costs and potentials information with the technological constraint (e.g., Färe et al. 2006; Murty et al. 2012; Wei et al. 2013; Du et al. 2015b; D'Inverno et al. 2018; Hu et al. 2020). First proposed by Shephard (1970), the distance function approach is able to evaluate the production efficiency with given technology. During the last two decades, researchers have tried to develop the original distance function approach by including pollution information. The revised versions of the distance function approach include Shephard input/output distance functions, directional output distance function and generalized non-radial directional distance function (Hailu and Veeman 2000; Färe et al. 2005, 2006; Areal et al. 2012; Tang et al. 2020). They have been employed to derive the shadow price of pollutant, representing the opportunity cost of reducing an incremental emission and could be interpreted as the marginal abatement cost (MAC). (e.g., Tang et al. 2016a, b; Yang et al. 2017; Singh and Gundimeda 2021; Zhang et al. 2021; Du et al. 2022). The derived shadow price might be applied to estimate the expected reduction attainable under a given emission charge, or likewise to determine the required emission charge for achieving a specific reduction target (Bento and Gianfrate 2020). Some have also used those distance functions to explore the reduction potential of pollution with given technological condition (Hampf and Krüger 2015; Zhang et al. 2016; Tang et al. 2020).

Compared with other versions of the distance function approach, the generalized non-radial directional distance function has its own merits. The main advantage is that it can reflect concurrent changes on inputs, outputs and pollutants emissions (Tang et al. 2020). In fact, the generalized non-radial directional distance function is a generalisation of Shephard input/output distance functions and directional output distance function. This chapter employs the generalized non-radial directional distance function to analyse carbon reduction costs, represented by shadow price, and potentials in China. Given that there are significant heterogeneities in economic, social, and natural conditions among provinces within the country, the analysis is therefore conducted at provincial level.

4.2 Methodology

4.2.1 The Parametric Generalized Non-Radial Directional Distance Function

Both nonparametric and parametric methods have been applied to estimate distance function. Specifically, data envelopment analysis (DEA) is a typical nonparametric approach that empirically quantifies the relative efficiency of decision-making units (DMUs), the homogeneous entities responsible for the conversion of inputs into outputs. Various types of DEA approach, such as the CCR (Charnes–Cooper–Rhodes) DEA model, the BCC (Banker–Charnes–Cooper) DEA model and the slacks-based measure (SBM) DEA model, are used to deal with pollutants in the total factor framework (Mehdiloozad et al. 2014; Sahoo et al. 2014; Halkos and Petrou 2019). A reason for the widely use of those types of DEA approach is their clear description as a movement along a given direction under the piece-wise-shaped technology curve. However, there have been some objections to the use of DEA approach in estimating distance function. The technology frontier estimated by DEA approach is piece-wise shaped, meaning that the frontier is indifferentiable. Besides, the estimated frontier is also sensitive to the outliers, influencing the result veracity (Hailu and Veeman 2001; Tang et al. 2016b; Yang et al. 2017). Those drawbacks can be well overcome by the parametric method, which allows for further hypotheses statistical testing and coefficient estimates (Khataza et al. 2019). Therefore, this chapter chooses to estimate distance function in a parametric way.

The production technology including polluting information is described as a set of all possible combinations of inputs $x = (x_1, \ldots, x_m) \in R_+$, products $o = (o_1, \ldots, o_n) \in R_+$ and GHG emissions $c \in R_+$. Hence, the production feasible set Ξ, a convex compact set, is defined as:

$$\Xi = \{(x, o, c) : x \text{ produces } (o, c)\}. \tag{4.1}$$

Ξ satisfies a serious of axioms, including null-jointness, possibility of inaction, no free lunch, free disposability of o, and strong disposability of x. Details about those axioms have been provided by Färe et al. (2005, 2006) and Tang et al. (2020).

The generalized non-radial directional distance function is then defined as follows:

$$\overrightarrow{D}_{GR}(x, o, c; d) = max\{\varphi : (x - \varphi d_x, o + \varphi d_o, c - \varphi d_c) \in \Xi, \varphi \in R_+\} \tag{4.2}$$

where $d = (d_x, d_o, d_c)$ represents the exogenously defined direction (Färe et al. 2010; Tang et al. 2020). φ represents the scaling term. $\overrightarrow{D}_{GR}$ aims at the concurrent maximal reduction of x and c and expansion of o within Ξ. If $\overrightarrow{D}_{GR} > 0$, technical inefficiency exists, saying that further reduction of o and c and expansion of x might achieve along the direction d. If $\overrightarrow{D}_{GR} = 0$, the production is thought to be fully efficient within Ξ. $\vec{D}_{GR}$ satisfies a series of commonly-used properties in the area of environmentally sensitive productivity analysis, such as weak disposability and translation properties. Readers are suggested to refer to, for instance, Murty et al. (2012), Wei et al. (2013) and Tang et al. (2021a), for the details of those properties.

4.2.2 Reduction Potentials and Costs

With given Ξ, the reduction potential of GHG emissions could be calculated as follows:

$$RP_c = \frac{c - [c - \vec{D}_{GR}(i, o, c; d)d_c]}{c} \times 100\% = \frac{\vec{D}_{GR}(i, o, c; d)d_c}{c} \times 100\% \tag{4.3}$$

where $c - [c - \vec{D}_{GR}(i, o, c; d)d_c]$ describes the maximum gainable amounts of GHG emissions within Ξ.

The profit function is then employed to estimate the shadow price of GHG emissions to represent the MAC (Tang et al. 2016b, 2021a; Färe et al. 2017). The estimated shadow price could be understood as the unit value of the GHG emissions produced at the tangency of the output-emission price-line and the technical frontier (Kuosmanen and Zhou 2021). In practice, it not only reflects the trade-off between the selected output and GHG emissions but also discloses the extent of producers' liability to society, the cost of generating one additional unit of GHG emissions at the technical frontier.

If the price of output o is known as p_o, then the shadow price of c could be calculated as follows:

$$p_c = -p_o \left(\frac{\partial \vec{D}_{GR}(i, o, c; d)}{\partial c} \Big/ \frac{\partial \vec{D}_{GR}(i, o, c; d)}{\partial o} \right). \tag{4.4}$$

4.2.3 Parametric and Estimation Specifications

The generalised quadratic form, suggested by Färe et al. (2010) and Tang et al. (2020), is chosen to parameterise the generalized non-radial directional distance function. Such a form is easy to apply and meets the requirements of globally imposing the translation property (Färe et al. 2010; Layer et al. 2020). For the purpose of making the parameterisation parsimonious, d is defined unitarily as $d = (1, 1, 1)$ to show the unit change (Khataza et al. 2019). Using mean-scaled data means that the mean values are adopted to determine the direction. Readers are suggested to refer to, for instance, Chambers et al. (2013), Layer et al. (2020) and Tang et al. (2021a), for the details of the generalised quadratic form.

To calculate the coefficients in the $\overrightarrow{D}_{GR}$, mathematical programming approach, rather than stochastic frontier approach, is applied in this study. Though both approaches are widely applied in related literature, mathematical programming approach has the advantages over stochastic frontier approach including less uncertain and lax distribution assumptions (Zhang and Choi 2014; Wang et al. 2018). Specifically, the mathematical programming approach used in this study aims at minimising the sum of the distance between the observations and the technical frontier constrained by feasibility, monotonicity, translation, and symmetry restrictions. The analysis adopts the APEAR package in R to estimate the coefficients in the generalized non-radial directional distance function.

4.2.4 Data Used

Given that there are significant heterogeneities in economic, social, and natural conditions among provinces within China, the analysis is conducted at provincial level. Hong Kong, Macau, Taiwan and Tibet are excluded because of data unavailability. The remaining provincial regions are grouped as eight economic areas to show regional conditions (Fig. 4.1).

China started its piloting regional carbon emission trading in 2013. To avoid potential interference by the emission trading in technical frontier and better show provincial carbon reduction costs and potentials within existing technologies, the study covers the 2003–2012 period. The study considers three inputs, one output and two pollutants. The inputs include in the parametric generalized non-radial directional distance function analysis are capital, labour and energy. The output is represented by provincial gross domestic product (GDP) at constant prices. Two pollutants are SO_2 emissions and CO_2 emissions. Since China launched strict regulation of SO_2 emissions in the study period that might influence production activities, SO_2 emissions are also included in the analysis. Inputs, output and SO_2 emissions data are obtained from *China Statistical Yearbook* (2004–2013), *China Environment Yearbook* (2004–2013) and *China Energy Yearbook* (2004–2013). CO_2 emissions are calculated using the IPCC approach (IPCC 2006).

Fig. 4.1 China's economic areas

4.3   Results and Discussion

4.3.1   Carbon Reduction Potentials

Overall, the estimated carbon reduction potentials fluctuated over time. For the whole country, the mean value of reduction potentials is 28.2%, implying that the country might achieve a 28% reduction in CO_2 emissions even without specific carbon regulations. Through improving technical efficiency, a massive reduction in emissions is feasible. Figure 4.2 shows that significant regional disparities exist. In general, south coast area has the smallest reduction potential. Within existing technologies, only a 4.1% reduction might be achieved in south coast area. For east coast area, the reduction potential reaches 12.2%. On the contrary, western and northern areas have much larger potentials. Among them, northwest and Middle Yellow River areas could reduce more than 45% carbon emissions. North coast, northeast and southwest areas are able to reduce more than 30% carbon emissions. The estimates argue that those western and northern areas could reduce their carbon emissions substantially within existing technical possibilities.

We try to explain the regional heterogeneity in terms of reduction potential. South coast and east coast areas are known as the country's economic powerhouse. They are large and fast-growing economies, mainly driven by manufacturing industries,

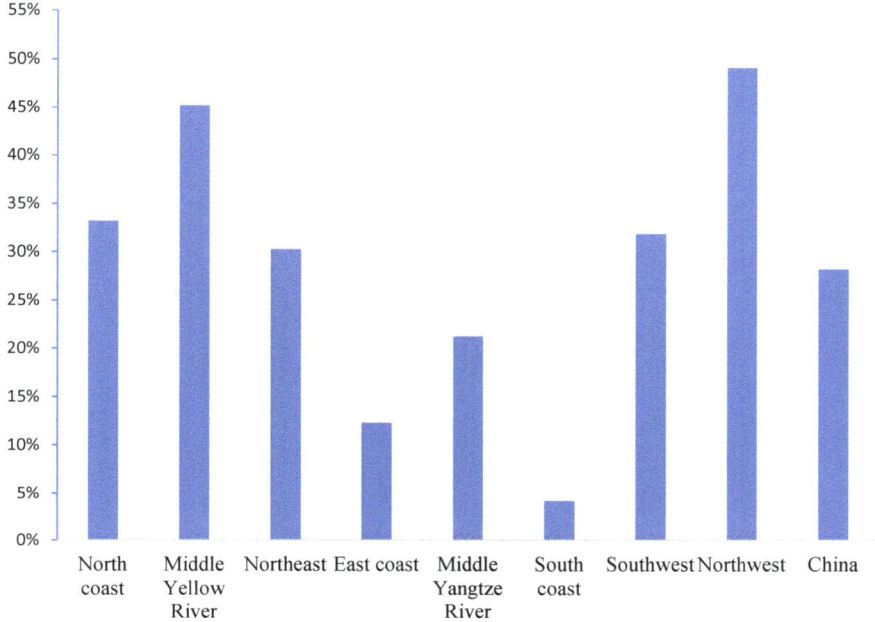

Fig. 4.2 Average carbon reduction potentials, 2003–2012

in terms of GDP. Those areas have been moving towards the development of high-tech, financial services and modern logistics industries over the last two decades. Compared with western and northern areas, south coast and east coast areas have higher efficiency level (Chen and Jia 2017; Yang and Zhang 2018; Wu et al. 2021a, b), thus contributing to smaller reduction potentials. In western and northern areas, many industries are still run in a traditional output-oriented way, characterized by high energy consumption and high carbon emissions, partly due to abundant local energy resources and relatively cheap energy prices. Besides, relatively low efficiency level also contributes to excessive carbon emissions (Tang et al. 2021b).

4.3.2 Carbon Reduction Costs

This analysis estimates shadow price of CO_2 to represent carbon reduction cost (Fig. 4.3). The estimated shadow price could be understood as the opportunity cost of reduction, reflecting the trade-off between the selected output and carbon emissions. It also demonstrates the extent of producers' liability to society, which describes the cost of generating one additional unit of carbon emissions at the technical frontier. Since the parametric generalized non-radial directional distance function approach

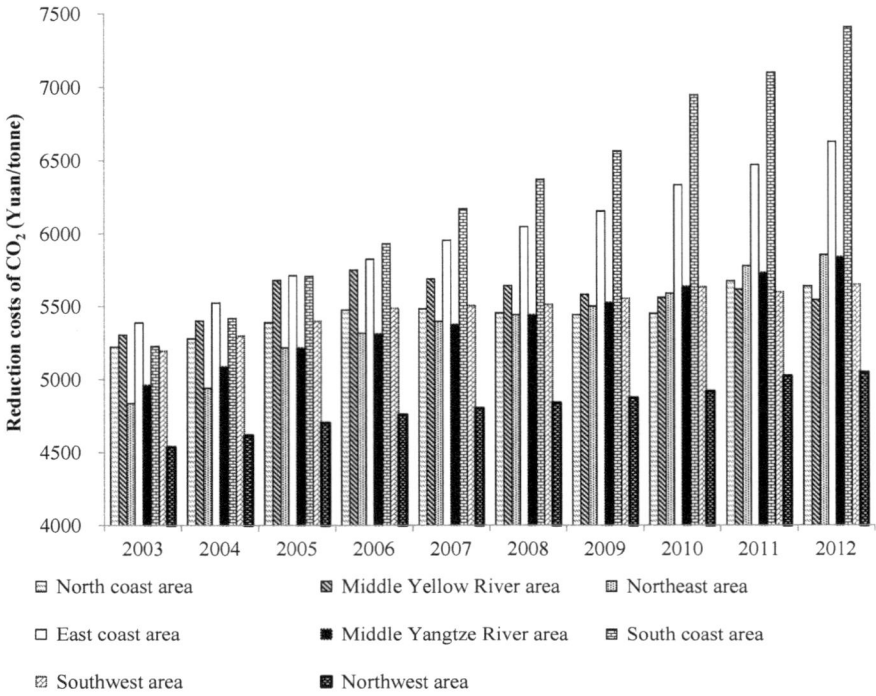

Fig. 4.3 Average carbon reduction costs, 2003–2012

used in this study does not consider possible use of new technologies, the estimated shadow price is better to be treated as a short-term partial equilibrium indicator.

In terms of the estimated shadow prices, a clear rising trend can be observed in the study period. The shadow value was about 5,500 Yuan/tonne on average across the country, which indicates the potential cost of reducing one additional unit of emissions with regard to corresponding less production of outputs at the technical frontier. For China's most economic areas, their average shadow prices of CO_2 went up steadily. However, the average shadow price in Middle Yellow River area increased at first and then turned to decline. Figure 4.3 shows that significant regional disparities in terms of CO_2 shadow prices exist. South coast and east coast areas had higher and faster-growing shadow prices, and their mean values exceeded 6,000 Yuan/tonne. On the contrary, northwest area had the lowest values continuously, with a mean value of 4,800 Yuan/tonne. Generally, other areas had shadow prices between 5,000 Yuan/tonne and 5,500 Yuan/tonne.

Some studies have noticed that there might be connection between MAC and emission intensities in China (Du et al. 2015a; Wu et al. 2019; Nakaishi 2021). They generally supported that lower emission intensities imply that further reduction will be more difficult, thus resulting in higher MAC. The results in Fig. 4.4 are generally in line with those studies. In China, south coast and east coast areas

Fig. 4.4 CO$_2$ shadow prices and carbon intensities, 2003–2012

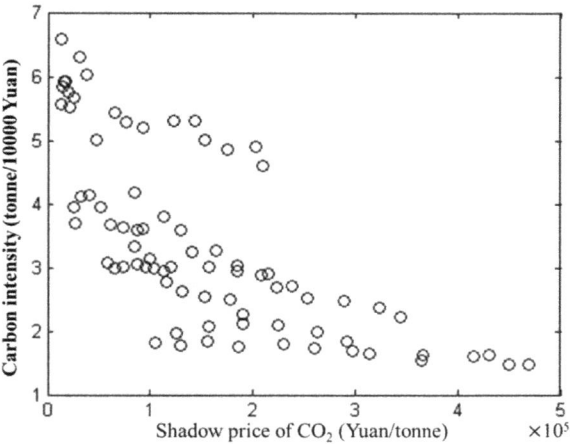

are large and fast-growing economies, who have been moving towards the development of high-tech, financial services and modern logistics industries over the last two decades. More tech- and capital-intensive industrial structure leads to lower emission intensity. Contrarily, northwestern provinces are still heavily relying on high-energy-consumption and high-carbon-emissions industries, such as coal-fired power, chemical and petrochemical, coal chemical, nonferrous metal, mining industries. Higher emission intensities mean that further reduction might be easier than south coast and east coast areas, which contributes to lower MAC in those regions.

Ma et al. (2019) critically reviewed economic studies on China's CO$_2$ MAC using distance function approach. They argued that the types of distance function used and the ways to estimate distance function may influence the empirical results. Our estimates are roughly in line with some of the reviewed studies using parametric directional distance function approaches (Figs. 4.5 and 4.6). It should be noted that the generalized non-radial directional distance function used in this study is a generalisation of Shephard input/output distance functions and directional output distance function. The distance approach used in this study might be more sufficient to reflect the reduction cost information since it is under the total-factor framework covering inputs, outputs and GHG emissions. Moreover, parametric estimation of distance function guarantees the differentiability of estimated frontier and is less likely to be influenced by outliers, compared with nonparametric methods.

China started its piloting regional carbon emission trading in 2013. Observed trading prices are between 10 Yuan/tonne and 100 Yuan/tonne, which are considerably lower than the results of this study. It should be noted that the short-term demand–supply relationship is the main driver of carbon trading prices (Creti et al. 2012; Tietenberg 2013; Zhao et al. 2018), while the reduction costs estimated in this study indicate the opportunity cost of reducing an incremental emission. Some have criticised China's carbon emission trading for being too generous with permits (Zhang et al. 2017; Shen and Wang 2019; Heggelund et al. 2022), which has inevitably

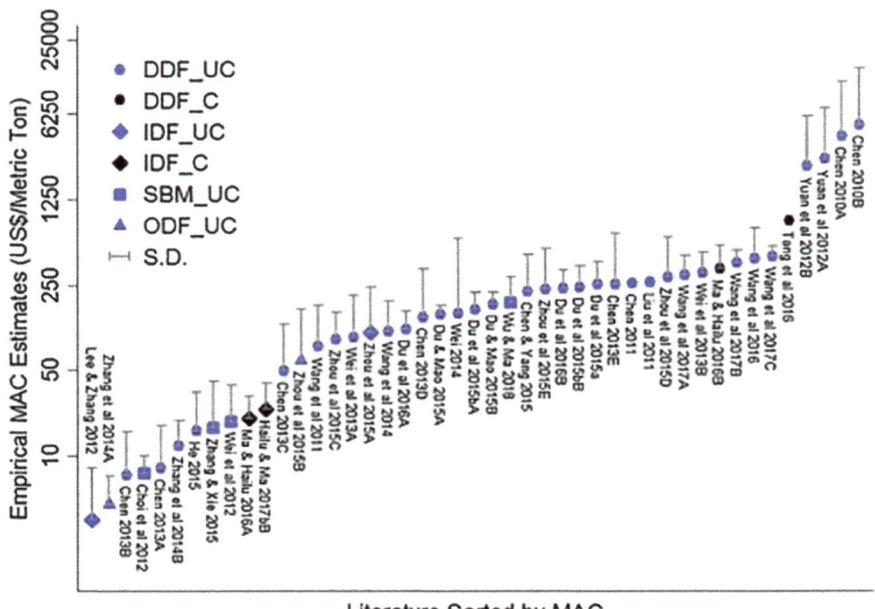

Fig. 4.5 Empirical MAC estimates of China's CO_2 emissions with various distance function approaches. *Notes* UC—unconditional MAC estimate; C—conditional MAC estimates; DDF—output directional distance function; IDF—input distance function; ODF—Shephard output distance function; SBM—slack based method; S.D.—standard deviations of MAC estimates; Figure source: Ma et al. (2019)

resulted in underpriced trading prices (Fig. 4.7). Measures are needed to rebalance the market with further refined trading system to effectively incentivise decarbonisation. For example, China's policymakers are encouraged to learn from the revised European Union's Emissions Trading System (EU ETS), which has reformed the rules to limit the oversupply permits. By doing so, the trading prices are likely to be booming. A more adequate carbon price will help internalize the external cost of climate change in the broadest possible range of economic decision-making and in setting economic incentives for decarbonized development. It can also help to mobilize the financial investments required to stimulate low-carbon technology and market innovation, thereby underpinning new, decarbonized drivers of development.

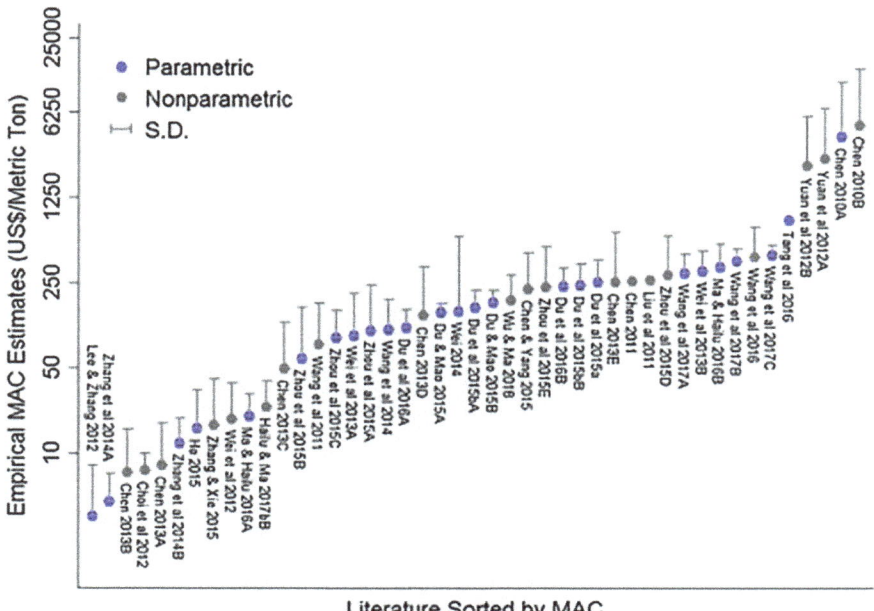

Fig. 4.6 Empirical MAC estimates of China's CO_2 Emissions, parametric and nonparametric approaches. *Notes* S.D.—standard deviations of MAC estimates; Figure source: Ma et al. (2019)

4.4 Conclusions

It is necessary to have a sound understanding of the costs and potentials associated with the carbon reduction efforts, which will contribute to designing and refining economic and climate policy regimes. This chapter employs the generalized non-radial directional distance function to analyse carbon reduction costs, represented by shadow price, and potentials in China at provincial level for the 2003–2012 period. The generalized non-radial directional distance function is parameterised using the generalised quadratic form. The coefficients in the distance function are estimated by applying mathematical programming approach.

The results show that the country might achieve a 28% reduction in CO_2 emissions even without specific carbon regulations. South coast and east coast areas have relatively small reduction potentials, while northwest and Middle Yellow River areas could reduce more than 45% carbon emissions. A clear rising trend can be observed in the study period in terms of the estimated shadow prices for most economic areas. The shadow value was about 5,500 Yuan/tonne on average across the country, more than 6,000 Yuan/tonne for south coast and east coast areas, and 4,800 Yuan/tonne for northwest area.

Readers need bear in mind that the approach adopted in this study does not consider possible use of new technologies, and therefore the estimated reduction potential

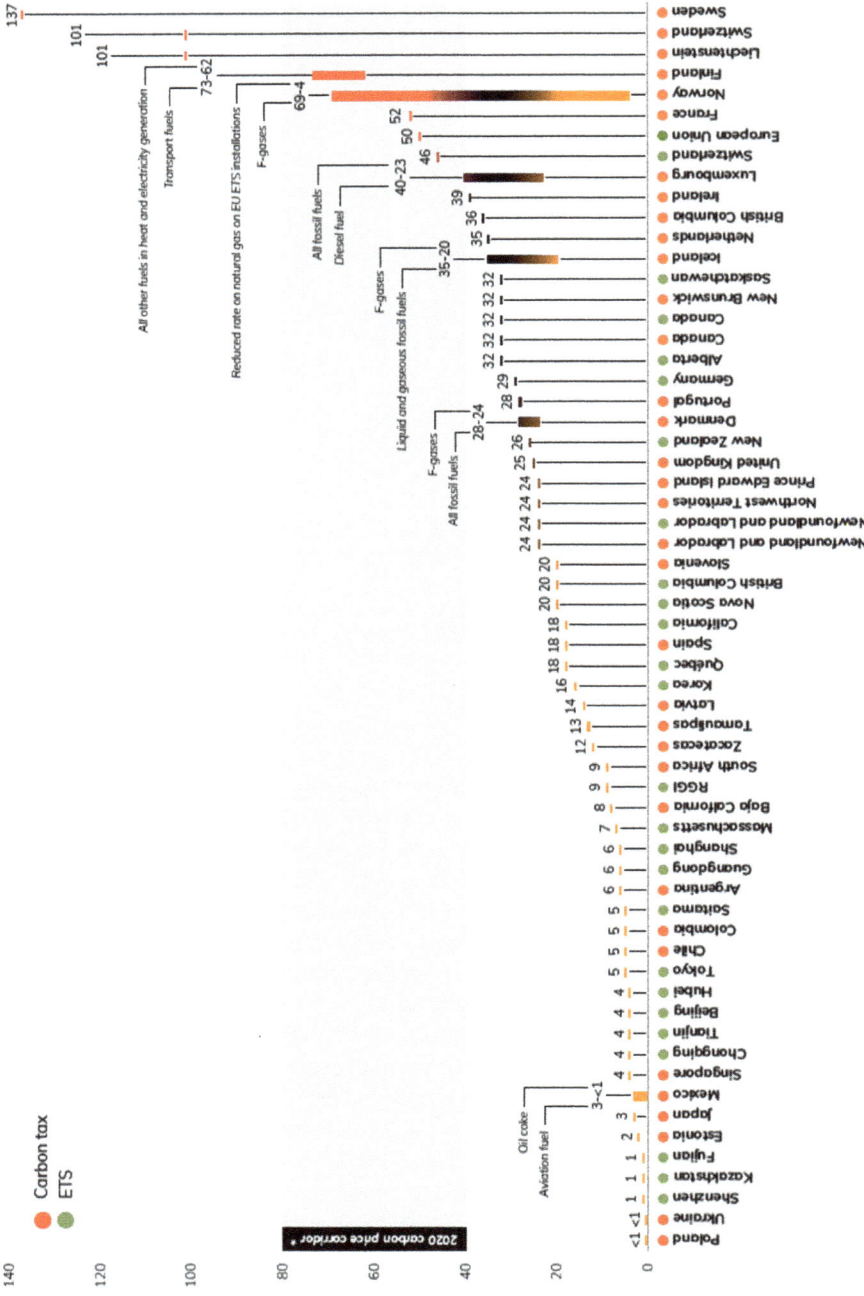

Fig. 4.7 Carbon prices as of April 1, 2021. *Source* World Bank Group (2021)

and cost results are better to be treated as short-term partial equilibrium indicators. Nevertheless, the empirical results will help policymakers, researchers and the general public to have a sound understanding of the costs and potential associated with the reduction efforts. Such an analysis could deliver rich empirical evidence to help design and refine China's net-zero emission policy regime.

References

Ai, H., Zhou, Z., Li, K., & Kang, Z. Y. (2021). Impacts of the desulfurization price subsidy policy on SO_2 reduction: Evidence from China's coal-fired power plants. *Energy Policy*, 157, 112477.

Antonakakis, N., Chatziantoniou, I., & Filis, G. (2017). Energy consumption, CO_2 emissions, and economic growth: An ethical dilemma. *Renewable and Sustainable Energy Reviews*, 68, 808-824.

Areal, F. J., Tiffin, R., & Balcombe, K. G. (2012). Provision of environmental output within a multi-output distance function approach. *Ecological Economics*, 78, 47-54.

Bento, N., Gianfrate, G., (2020). Determinants of internal carbon pricing. *Energy Policy*, 143, 111499.

Canadell, J. G., Le Quéré, C., Raupach, M. R., Field, C. B., Buitenhuis, E. T., Ciais, P., Conway, T.J., Gillett, N.P., Houghton, R.A., & Marland, G. (2007). Contributions to accelerating atmospheric CO_2 growth from economic activity, carbon intensity, and efficiency of natural sinks. *Proceedings of the National Academy of Sciences*, 104(47), 18866-18870.

Chambers, R., Färe, R., Grosskopf, S., & Vardanyan, M. (2013). Generalized quadratic revenue functions. *Journal of Econometrics*, 173(1), 11-21.

Chen, L., & Jia, G. (2017). Environmental efficiency analysis of China's regional industry: A data envelopment analysis (DEA) based approach. Journal of Cleaner Production, 142, 846-853.

Chen, X., Chen, G., Lin, M., Tang, K., & Ye, B. (2022). How does anti-corruption affect enterprise green innovation in China's energy-intensive industries?. *Environmental Geochemistry and Health*, 44, 2919–2942.

Creti, A., Jouvet, P. A., & Mignon, V. (2012). Carbon price drivers: Phase I versus Phase II equilibrium? *Energy Economics*, 34(1), 327-334.

D'Inverno, G., Carosi, L., Romano, G., & Guerrini, A. (2018). Water pollution in wastewater treatment plants: An efficiency analysis with undesirable output. *European Journal of Operational Research*, 269(1), 24-34.

Doelman, J. C., Stehfest, E., van Vuuren, D. P., Tabeau, A., Hof, A. F., Braakhekke, M. C., Gernaat, D.E., van den Berg, M., van Zeist, W.J., Daioglou, V., van Meijl, H., & Lucas, P. L. (2020). Afforestation for climate change mitigation: Potentials, risks and trade-offs. *Global Change Biology*, 26(3), 1576-1591.

Du, L., Hanley, A., & Wei, C. (2015a). Estimating the marginal abatement cost curve of CO_2 emissions in China: Provincial panel data analysis. *Energy Economics*, 48, 217-229.

Du, L., Hanley, A., & Wei, C. (2015b). Marginal abatement costs of carbon dioxide emissions in China: A parametric analysis. *Environmental and Resource Economics*, 61(2), 191-216.

Du, L., Lu, Y., & Ma, C. (2022). Carbon efficiency and abatement cost of China's coal-fired power plants. *Technological Forecasting and Social Change*, 175, 121421.

Färe, R., Grosskopf, S., Noh, D.W., Weber, W. (2005). Characteristics of a polluting technology: Theory and practice. *Journal of Econometrics*, 126(2), 469–492.

Färe, R., Grosskopf, S., Weber, W. L. (2006). Shadow prices and pollution costs in US agriculture. *Ecological Economics*, 56(1), 89-103.

Färe, R., Martins-Filho, C., Vardanyan, M. (2010). On functional form representation of multi–output production technologies. *Journal of Productivity Analysis*, 33(2), 81–96.

Färe, R., Pasurka, C., & Vardanyan, M. (2017). On endogenizing direction vectors in parametric directional distance function-based models. *European Journal of Operational Research*, 262(1), 361-369.

Hailu, A., Veeman, T. S. (2000). Environmentally sensitive productivity analysis of the Canadian pulp and paper industry, 1959–1994: An input distance function approach. *Journal of Environmental Economics and Management*, 40(3), 251–274.

Hailu, A., & Veeman, T. S. (2001). Non-parametric productivity analysis with undesirable outputs: An application to the Canadian pulp and paper industry. *American Journal of Agricultural Economics*, 83(3), 605-616.

Halkos, G., & Petrou, K. N. (2019). Treating undesirable outputs in DEA: A critical review. *Economic Analysis and Policy*, 62, 97-104.

Hampf, B., & Krüger, J. J. (2015). Optimal directions for directional distance functions: An exploration of potential reductions of greenhouse gases. *American Journal of Agricultural Economics*, 97(3), 920-938.

Heggelund, G., Stensdal, I., & Duan, M. (2022). China's carbon market: Potential for success?. *Politics and Governance*, 10(1), 265-274.

Hu, Y., Ren, S., Wang, Y., & Chen, X. (2020). Can carbon emission trading scheme achieve energy conservation and emission reduction? Evidence from the industrial sector in China. *Energy Economics*, 85, 104590.

IPCC. (2006). *Greenhouse Gas Inventory: IPCC Guidelines for National Greenhouse Gas Inventories*. United Kingdom Meteorological Office, Bracknell, England.

Kaufmann, R. K., Davidsdottir, B., Garnham, S., & Pauly, P. (1998). The determinants of atmospheric SO_2 concentrations: Reconsidering the environmental Kuznets curve. *Ecological Economics*, 25(2), 209-220.

Khataza, R. R., Hailu, A., Doole, G. J., Kragt, M. E., & Alene, A. D. (2019). Examining the relationship between farm size and productive efficiency: A Bayesian directional distance function approach. *Agricultural Economics*, 50(2), 237-246.

Kuosmanen, T., Zhou, X. (2021). Shadow prices and marginal abatement costs: Convex quantile regression approach. *European Journal of Operational Research*, 289(2), 666-675.

Layer, K., Johnson, A. L., Sickles, R. C., & Ferrier, G. D. (2020). Direction selection in stochastic directional distance functions. *European Journal of Operational Research*, 280(1), 351-364.

Liu, D., Ren, S., & Li, W. (2022). SO_2 emissions trading and firm exports in China. *Energy Economics*, 109, 105978.

Ma, C., Hailu, A., & You, C. (2019). A critical review of distance function based economic research on China's marginal abatement cost of carbon dioxide emissions. *Energy Economics*, 84, 104533.

Mehdiloozad, M., Sahoo, B. K., & Roshdi, I. (2014). A generalized multiplicative directional distance function for efficiency measurement in DEA. *European Journal of Operational Research*, 232(3), 679-688.

Murty, S., Russell, R. R., & Levkoff, S. B. (2012). On modeling pollution-generating technologies. *Journal of Environmental Economics and Management*, 64(1), 117-135.

Nakaishi, T. (2021). Developing effective CO_2 and SO_2 mitigation strategy based on marginal abatement costs of coal-fired power plants in China. *Applied Energy*, 294, 116978.

Rosenzweig, C., Mbow, C., Barioni, L. G., Benton, T. G., Herrero, M., Krishnapillai, M., Liwenga, E.T., Pradhan, P., Rivera-Ferre, M.G., Sapkota, T., & Portugal-Pereira, J. (2020). Climate change responses benefit from a global food system approach. *Nature Food*, 1(2), 94-97.

Sahoo, B. K., Mehdiloozad, M., & Tone, K. (2014). Cost, revenue and profit efficiency measurement in DEA: A directional distance function approach. *European Journal of Operational Research*, 237(3), 921-931.

Sarkodie, S. A., & Strezov, V. (2018). Empirical study of the environmental Kuznets curve and environmental sustainability curve hypothesis for Australia, China, Ghana and USA. *Journal of Cleaner Production*, 201, 98-110.

Shen, W., & Wang, Y. (2019). Adaptive policy innovations and the construction of emission trading schemes in China: Taking stock and looking forward. *Environmental Innovation and Societal Transitions*, 30, 59-68.

Shephard, R. W. (1970). *Theory of Cost and Production Functions*. Princeton University Press, Princeton.

Singh, A., & Gundimeda, H. (2021). Measuring technical efficiency and shadow price of water pollutants for the leather industry in India: A directional distance function approach. *Journal of Regulatory Economics*, 59(1), 71-93.

Tang, K., & Ma, C. (2022). The cost-effectiveness of agricultural greenhouse gas reduction under diverse carbon policies in China. *China Agricultural Economic Review*. https://doi.org/10.1108/CAER-01-2022-0008

Tang, K., Gong, C., & Wang, D. (2016a). Reduction potential, shadow prices, and pollution costs of agricultural pollutants in China. *Science of the Total Environment*, 541, 42-50.

Tang, K., Hailu, A., Kragt, M. E., & Ma, C. (2016b). Marginal abatement costs of greenhouse gas emissions: Broadacre farming in the Great Southern Region of Western Australia. *Australian Journal of Agricultural and Resource Economics*, 60(3), 459-475.

Tang, K., Hailu, A., Kragt, M. E., & Ma, C. (2018). The response of broadacre mixed crop-livestock farmers to agricultural greenhouse gas abatement incentives. *Agricultural Systems*, 160, 11-20.

Tang, K., He, C., Ma, C., & Wang, D. (2019). Does carbon farming provide a cost-effective option to mitigate GHG emissions? Evidence from China. *Australian Journal of Agricultural and Resource Economics*, 63(3), 575-592.

Tang, K., Hailu, A., & Yang, Y. (2020). Agricultural chemical oxygen demand mitigation under various policies in China: A scenario analysis. *Journal of Cleaner Production*, 250, 119513.

Tang, K., Wang, M., & Zhou, D. (2021a). Abatement potential and cost of agricultural greenhouse gases in Australian dryland farming system. *Environmental Science and Pollution Research*, 28(17), 21862-21873.

Tang, K., Xiong, C., Wang, Y., & Zhou, D. (2021b). Carbon emissions performance trend across Chinese cities: Evidence from efficiency and convergence evaluation. *Environmental Science and Pollution Research*, 28(2), 1533-1544.

Tang, K., Zhou, Y., Liang, X., & Zhou, D. (2021c). The effectiveness and heterogeneity of carbon emissions trading scheme in China. *Environmental Science and Pollution Research*, 28(14), 17306-17318.

Tietenberg, T. H. (2013). Reflections—carbon pricing in practice. *Review of Environmental Economics and Policy*, 7(2), 313–329.

Wang, K., Yang, K., Wei, Y. M., & Zhang, C. (2018). Shadow prices of direct and overall carbon emissions in China's construction industry: a parametric directional distance function-based sensitive estimation. *Structural Change and Economic Dynamics*, 47, 180-193.

Wei, C., Löschel, A., & Liu, B. (2013). An empirical analysis of the CO_2 shadow price in Chinese thermal power enterprises. *Energy Economics*, 40, 22-31.

World Bank Group. (2021) State and Trends of Carbon Pricing 2021. Washinton, DC: World Bank. https://openknowledge.worldbank.org/handle/10986/35620

Wu, J., Ma, C., & Tang, K. (2019). The static and dynamic heterogeneity and determinants of marginal abatement cost of CO_2 emissions in Chinese cities. *Energy*, 178, 685-694.

Wu, J., Xu, H., & Tang, K. (2021a). Industrial agglomeration, CO_2 emissions and regional development programs: A decomposition analysis based on 286 Chinese cities. *Energy*, 225, 120239.

Wu, J., Feng, Z., & Tang, K. (2021b). The dynamics and drivers of environmental performance in Chinese cities: A decomposition analysis. *Environmental Science and Pollution Research*, 28(24), 30626-30641.

Yang, L., Tang, K., Wang, Z., An, H., & Fang, W. (2017). Regional eco-efficiency and pollutants' marginal abatement costs in China: A parametric approach. *Journal of Cleaner Production*, 167, 619-629.

Yang, L., & Zhang, X. (2018). Assessing regional eco-efficiency from the perspective of resource, environmental and economic performance in China: A bootstrapping approach in global data envelopment analysis. *Journal of Cleaner Production, 173*, 100-111.

Yuan, F., Tang, K., & Shi, Q. (2021). Does Internet use reduce chemical fertilizer use? Evidence from rural households in China. *Environmental Science and Pollution Research, 28*(5), 6005-6017.

Yuan, F., Tang, K., Shi, Q., Qiu, W., & Wang, M. (2022). Rural women and chemical fertiliser use in rural China. *Journal of Cleaner Production, 344*, 130959.

Zhang, J., Wang, Z., & Du, X. (2017). Lessons learned from China's regional carbon market pilots. *Economics of Energy & Environmental Policy, 6*(2), 19-38.

Zhang, N., & Choi, Y. (2014). A note on the evolution of directional distance function and its development in energy and environmental studies 1997–2013. *Renewable and Sustainable Energy Reviews, 33*, 50-59.

Zhang, N., Huang, X., & Liu, Y. (2021). The cost of low-carbon transition for China's coal-fired power plants: A quantile frontier approach. *Technological Forecasting and Social Change, 169*, 120809.

Zhang, Y. J., Hao, J. F., & Song, J. (2016). The CO_2 emission efficiency, reduction potential and spatial clustering in China's industry: Evidence from the regional level. *Applied Energy, 174*, 213-223.

Zhao, X., Han, M., Ding, L., & Kang, W. (2018). Usefulness of economic and energy data at different frequencies for carbon price forecasting in the EU ETS. *Applied Energy, 216*, 132-141.

Chapter 5
Urban Carbon Reduction Costs and Potentials in China: A Nonparametric Approach

Jianxin Wu and Kai Tang

5.1 Introduction

China's ongoing rapid urbanization and industrialization lead to growing demand for energy, which might boost related carbon emissions (Zhu and Peng 2012; Chen et al. 2020; Wu et al. 2021a, b; Lv et al. 2022). During the last several decades, the country's growth has been mainly driven by energy-intensive production and infrastructure construction, and the country's production sectors have been shifted from less to more energy-intensive consumption (U.S. Energy Information Administration 2018; Wu et al. 2019a, b; Chen et al. 2022). In addition, rapid urbanization is thought to increase energy consumption due to the increasing levels of housing rates, mobility and heat island effect (Shahbaz et al. 2016; Mohammadi and Taylor 2017; Li and Sun 2018; Xu et al. 2021). Empirical evidence shows that for every 1% increase in economic urbanization, carbon emissions will increase by more than 1% (Lv et al. 2022). Consequently, massive carbon emissions are produced. In 2019, China emitted 27% of the world's GHG, while the US and India contributed 11% and 6%, respectively. GHG emissions from all members of the Organization for Economic Cooperation and Development (OECD) as well as all 27 EU member states were 14,057 MMt CO_2e, while emissions in China were 14,093 MMt CO_2e (Larsen et al. 2021).

China's current administration has recognized its role in global fight against climate change and set carbon reduction and climate change mitigation as its top priority. National targets for carbon intensity reductions have been unveiled in the

J. Wu
Institute of Resource, Environment and Sustainable Development, School of Economics, Jinan University, Guangzhou 510632, China

K. Tang (✉)
School of Economics and Trade, Guangdong University of Foreign Studies, Guangzhou 510006, China
e-mail: francistang1988@hotmail.com

11th Five Year Plan (FYP) (2006–2010), the 12th FYP (2011–2015), the 13th FYP (2016–2020) and the recently released 14th FYP (2021–2025). Recently, a series of 2030 pledges have been made, including increasing the share of non-fossil fuels in primary energy consumption mix to 25%, decreasing carbon intensity, the amount of CO_2 emissions per unit of GDP, by 65%, increasing the combined capacity of solar and winder power generators to 1.2 billion kilowatts, and enhancing the forest stock volume by 6 billion cubic meters from the 2005 level. Those efforts so far have been made mostly by implementing command-and-control regulations, which might not be as efficient as expected (Kostka 2016; Chang et al. 2022; Tang and Ma 2022).

Besides, market-based regulations, which are widely believed to be more efficient and flexible than command-and-control regulations (Tang et al. 2019; Wu et al. 2019a, b), have also been introduced. China's policymakers have also been encouraged to use market-based regulations to achieve their goal of peaking carbon emissions by 2030 and reaching net zero emissions by 2060. Since 2013, eight pilot carbon trading markets in Beijing, Tianjin, Shanghai, Hubei, Guangdong, Chongqing, Shenzhen, and Fujian have been established. A national carbon trading market, which is the world's biggest, was officially launched in July 2021. The newly established market covers 2,225 domestic power operators using fossil fuels, accounting for more than 40% of the country's total CO_2 emissions and 14% of fossil-fuel-related CO_2 emissions in the world. Nevertheless, the country's carbon trading market, hindered by low prices and thin trading, is still criticized for its limited coverage and oversupply (Cui et al. 2021; Tang et al. 2021d; Liao and Yao 2022).

To refine China's climate change policy regime, empirical evidence on carbon emissions profile, including reduction costs and potentials, is much needed. In terms of empirical studies on China's carbon reduction costs and potentials, most have been conducted at provincial level (e.g., Guo et al. 2011; Tang et al. 2016; Yang et al. 2017; Wu et al. 2019a, b, 2020; He et al. 2020) or industrial level (e.g., Lin and Moubarak 2014; Yang et al. 2018; Tang et al. 2019; Du et al. 2021). Considering the close relationship between urbanization and carbon emissions and cities are responsible for the majority of carbon emissions in China (Dhakal 2009; Chen et al. 2017, 2021; Tang et al. 2021a, c), it is therefore necessary to understand the country's carbon emissions profile at city level, which could provide basic information to achieve the country's climate change goals. Moreover, the city-level analysis has an advantage of addressing intra-provincial heterogeneity while still delivering economy-wide evidence in terms of carbon reduction costs and potentials. Recently, some began to address this issue at city level. Wang et al. (2017) proposed an improved slacks-based measure approach and applied it to analyse the carbon reduction costs and potentials of China's 285 cities for the 2008–2012 period. They argued that eastern cities are encouraged to reduce more carbon emissions with the considerations of both reduction costs and potentials. A similar conclusion was drawn by Zhou et al. (2018b) who adopted the non-parametric non-radial directional distance function approach to explore the carbon reduction costs and potentials of China's 71 cities for the 2005–2012 period. Partially due to the data availability issue of city-level carbon emissions, empirical studies on investigating China's city-level carbon reduction costs and potentials are still rare.

There are generally two ways of estimating carbon reduction costs and potentials. The first is to use integrated assessment models to explore predicted scenarios. The second is to apply distance function methods to analyse real-world datasets. The results of the first type evaluations are typically derived as part of a national or even global forecasting modelling, and generally provide little information at the disaggregated local levels at which China's current carbon reduction regulations are implemented (Hailu and Ma 2017). In terms of the second way, radial approaches including Shephard input/output distance functions and non-radial approaches have been applied, and non-radial approaches are thought to be more flexible and accurate metrics with pollutants (Vardanyan and Noh 2006; Wu et al. 2019a, b; Tang et al. 2021b). In literature, the second type usually uses the shadow prices of carbon, derived from the proposed distance function, to represent carbon reduction costs.

This chapter employs a non-radial approach, the slacks-based measure data envelopment analysis (SBM-DEA) method, to analyse city-level carbon reduction costs and potentials in China. A dataset including 286 prefectural-and-above (PAA) cities has been complied. Carbon reduction costs are estimated through measuring shadow prices. Such an analysis helps us to understand China's carbon emissions profile at city level and address intra-provincial heterogeneity while still delivering economy-wide evidence in terms of carbon reduction costs and potentials.

5.2   Methodology

5.2.1   Estimating Carbon Reduction Potentials

The SBM method, originally proposed by Tone (2001), is a widely applied non-radial approach which could adjust each input and output separately (Gómez-Calvet et al. 2014; Aparicio et al. 2017; Lee 2021; Meng and Pang 2022). The SBM-DEA method, which adds the polluting information into both the objective function and constraint function, is a revised version of Tone (2001)'s SBM method (Chang et al. 2013; Sueyoshi et al. 2017; Zhou et al. 2018a). This chapter also employs the non-radial SBM-DEA method to analyse city-level carbon reduction costs and potentials in China.

The production technology including polluting information for J cities is described as a set of all possible combinations of inputs $x = (x_1, \ldots, x_m) \in R_+^M$, products $o = (o_1, \ldots, o_n) \in R_+^N$ and emissions of pollutants (e.g., GHG emissions) $c = (c_1, \ldots, c_l) \in R_+^L$. Hence, the production feasible set Ξ is defined as:

$$\Xi = \{(x, o, c) : x \geq X\lambda, o \leq O\lambda, c \geq C\lambda, \lambda \geq 0\}. \tag{5.1}$$

where $\lambda \in R^J$ is an intensity vector, implying that Ξ satisfies the constant returns to scale assumption.

The non-radial SBM-DEA model (Chang et al. 2013) is defined as follows:

$$\sigma^* = \min \frac{1 - \frac{1}{M}\sum_{m=1}^{M}\frac{s_m^x}{x_{m,j}}}{1 + \frac{1}{N+L}\left(\sum_{n=1}^{N}\frac{s_n^o}{o_{n,j}} + \sum_{l=1}^{L}\frac{s_l^c}{c_{l,j}}\right)}, \sigma^* \in [0, 1]$$

$$S.T.$$

$$x_{j,m} = X\lambda + s_m^x, \forall m; \quad o_{j,n} = O\lambda - s_n^o, \forall n; \quad c_{j,l} = C\lambda + s_l^c, \forall l;$$

$$s_m^x \geq 0, \ s_n^o \geq 0, \ s_l^c \geq 0, \ \lambda \geq 0$$

(5.2)

where s_m^x, s_n^o and s_l^c are the slack variables vectors. Among them, s_m^x and s_l^c denote excesses in x and c, respectively, whereas s_n^o represents the shortage of o. If $\sigma^* = 1$, the corresponding decision making unit (x_j, o_j, c_j) is efficient, and all the corresponding slack variables are 0. In practice, σ^* might be understood as urban total factor environmental efficiency.

Specifically, the factor efficiency of pollutant can be defined as:

$$E_{lc} = 1 - \frac{s_l^c}{c_{l,j}}$$

(5.3)

The estimated factor efficiency of pollutant reflects reduction potential. A higher value of E_{lc} indicates a lower reduction potential level, while a lower value of E_{lc} implies a higher reduction potential level.

5.2.2 Estimating Carbon Reduction Costs

Following Choi et al. (2012) and Kuhn et al. (2020), the carbon reduction costs are estimated as follows:

$$\text{Max} \theta^o o_j - \theta^x x_j - \theta^c c_j$$

$$S.T.$$

$$\theta^o O - \theta^x X - \theta^c C \leq 0$$

$$\theta^x > \frac{1}{n}\left(\frac{1}{x_j}\right)$$

$$\theta^o \geq \frac{1 + \theta^o o_j - \theta^x x_j - \theta^c c_j}{N+L}\left(\frac{1}{o_j}\right)$$

$$\theta^c \geq \frac{1 + \theta^o o_j - \theta^x x_j - \theta^c c_j}{N+L}\left(\frac{1}{c_j}\right)$$

(5.4)

where, non-negative vectors θ^x, θ^o and θ^c are virtual prices of x, o and c, respectively. Then, the shadow price of pollutant c_l is defined as:

$$p^c = p^y \times \frac{\theta^c}{\theta^o}.$$

5.2.3   Variables and Data

In China's mainland, there were 288 PAA cities by 2013. However, data for two Tibetan cities Lasa and Rikaze is largely unavailable. Therefore, this chapter compiles a panel dataset of 286 PAA cities for the 2002–2013 period and sets 2002 as the base year. Given that there are significant heterogeneities in economic, social, and natural conditions among regions within the country, cities are grouped into four groups according to the official classification of the National Bureau of Statistics of China as follows: the eastern cities include those in Beijing, Tianjin, Hebei, Shanghai, Jiangsu, Zhejiang, Fujian, Shandong, Guangdong and Hainan; the central cities include those in Shanxi, Anhui, Jiangxi, Henan, Hubei, and Hunan; the western cities include those in Inner Mongolia, Guangxi, Chongqing, Sichuan, Guizhou, Yunnan, Tibet, Shaanxi, Gansu, Qinghai, Ningxia, and Xinjiang; and the northeast cities include those in Liaoning, Jilin and Heilongjiang.

Inputs include labour, capital and energy. Output is represented by gross city product (GCP). Pollutant includes CO_2 emissions. The total number of employees is selected to denote labour. Data for gross city product is measured at 2002 constant price. Capital is estimated using the perpetual inventory approach (Wu 2009; Wu and Ma 2019). Key energy consumption sources, electricity, coal gas and liquefied petroleum gas, transportation, and heating, are included (Glaeser and Kahn 2010; Wu and Ma 2019). Readers are referred to Wu and Ma (2019) to obtain the details of the energy consumption calculation approach used. CO_2 emissions are calculated using the accounting approach suggested by IPCC (2006). The required data information for compiling the dataset is drawn from *China City Statistical Yearbook* (NBSC 2003–2014a), *China Urban Construction Statistical Yearbook* (NBSC 2003–2014b), *Chinese Statistical Yearbook for Regional Economy* (NBSC 2003–2014c) and *China Statistical Yearbook* (NBSC 2003–2014d).

5.3 Results and Discussion

5.3.1 Estimates of Urban Carbon Reduction Potentials

The estimated results show that the factor efficiency of CO_2 emissions generally gained slow improvement for the 2002–2013 period, indicating that urban carbon reduction potential level showed a slightly decreasing trend. In average, the eastern cities had the highest value (about 0.5), whereas the northeastern cities had the lowest value (about 0.3). The results of the central and western cities are quite similar (about 0.4). These results reveal that the eastern cities had smaller carbon reduction potential, while the northeastern cities had larger reduction potential.

We try to understand the above results from diverse perspectives. China's northeastern cities have bitterly cold harsh winters, with temperatures as low as −40 °C. The winter heating season in northeastern cities normally lasts from October to mid-April or even May. The winter heating system is basically coal-based and thought to be less efficient in energy use compared to the heating system in developed economies (Almond et al. 2009). Due to the climate and technological characteristics, winter heating in the northeastern cities accounts for a large share of local energy consumption, thus causing enormous carbon emissions (Liu et al. 2019; Shao and Jin 2020). Besides, the northeastern region is the old heavy industry centre of the country. Mao Zedong, with the aid of the Soviet Union, made the northeastern the centre of heavy industry. The local economy has long been driven by heavy industry, e.g., steelmaking, mining, chemical, oil and gas sectors, which are usually characterized as high energy consumption and high carbon emissions. Consequently, both the average per capita carbon emissions and carbon emission intensity in those northeastern cities are higher than the national levels (Liu et al. 2019). Zheng et al. (2011) found that relative to the average city household, carbon emissions are 69% higher in the northeastern region. Hence, it is not surprising that the northeastern cities had larger reduction potential.

Compared with the northeastern cities, the eastern cities have much shorter and milder winters. Since 1950s, the Chinese government has provided winter central heating for the cities north of the line of Qinling Mountains and Huai River, where the daily average temperatures are lower than 5 °C for more than 90 days yearly.[1] Residents in most eastern cities do not have access to central heating due to China's central heating policy, thus limiting energy use and the corresponding carbon emissions. Moreover, the eastern cities generally have greener industry than the northeastern cities. Relatively high economic, social and technological development level also help the eastern cities to promote the sustainable transformation of local economic and social system (Wang et al. 2022). Those may contribute to the relatively smaller carbon reduction potential of the eastern cities.

---

[1] https://news.cgtn.com/news/2020-11-19/Will-the-debate-over-winter-heating-systems-in-China-continue--VwGcpFqDbq/index.html.

City size in terms of urban population might also have an impact on carbon emissions profile considering that it can influence economic progress, technological level and energy use (Henderson 2010; Zarco-Soto et al. 2021). This chapter divides sample cities into three roughly equal groups according to population size, including 95 large cities, 95 medium-sized cities and 96 small cities. The results show that large cities tended to have higher CO_2 emissions efficiency, whereas small cities had lower efficiency. This indicates that large cities had smaller carbon reduction potential, while small cities had larger reduction potential. The result is broadly consistent with finding of Wu et al. (2017).

5.3.2 Estimates of Urban Carbon Reduction Costs

Figures 5.1 and 5.2 report the estimated average carbon reduction costs for PAA cities, represented by shadow prices. Nationally, the average and median values were about 1,190 Yuan/tonne and 1,050 Yuan/tonne, respectively. These results are considerably lower than the provincial estimates in Chap. 4. Some of the differences between our estimates and those of Chap. 4 may stem from a difference in datasets studied. Moreover, Chap. 4 employs a parametric non-radial directional distance function approach while this chapter uses a non-parametric one. Such a difference may also cause variability on estimates (Zhou et al. 2014; Hailu and Ma 2017).

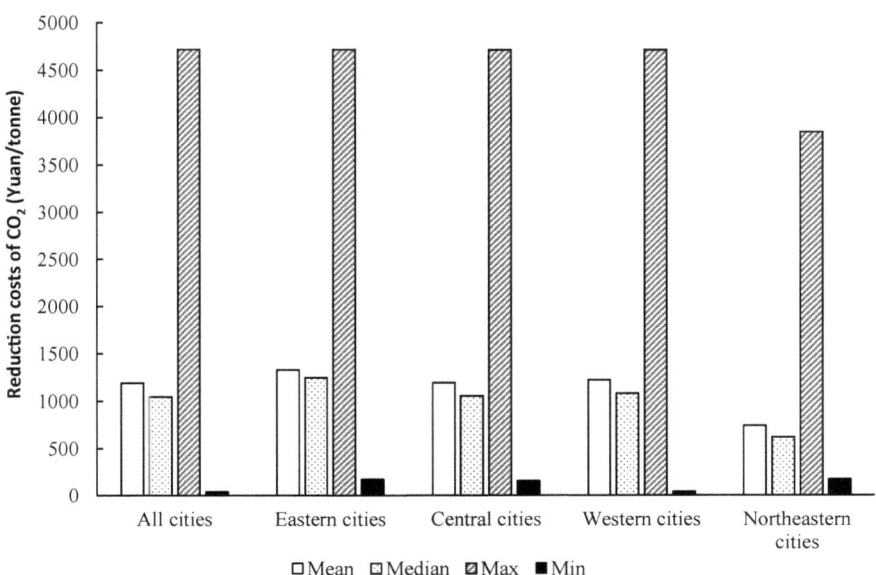

Fig. 5.1 Urban carbon reduction costs by region

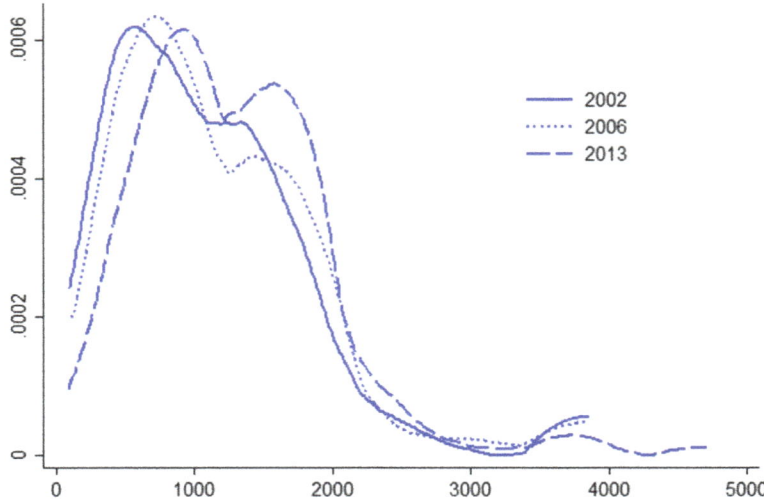

Fig. 5.2 Distributions of carbon reduction costs in selected years

Figures 5.1 and 5.2 also show substantially geographical heterogeneity across regions. Such a heterogeneity demonstrates that the newly established national carbon trading market is likely to generate more positive externalities than command-and-control regulations. Averagely, the eastern cities had the highest carbon reduction cost whereas the northeastern cities had the lowest reduction cost. The carbon reduction cost of the eastern cities was slightly higher than those of the central and western cities. As we have stated above, both the average per capita carbon emissions and carbon emission intensity in the northeastern cities are higher than the national levels due to their heavy-industry-driven economies and low efficient winter heating. These might result in the relatively low carbon reduction cost of the northeastern cities. Among all PAA cities, Shizhuishan had the lowest reduction cost (about 120 Yuan/tonne). This city is known as the "western coal city" since it produces high-quality anthracite. Local economy has long been highly dominated by coal mining, coking, and metallurgy industries. This city is also among the world's most polluted cities.[2] It is therefore not surprising that Shizhuishan had the low reduction cost.

The estimates show that the annual average carbon reduction costs for the eastern, central, and western cities fluctuated during the studied period. However, the northeatern cities experienced a considerable increase of 94.9%. Figure 5.3 plots the kernel density distribution of carbon reduction costs for the PAA cities in selected years. It can be observed that the distributions were right-skewed and multimodal. The movement of distributions also confirmed an overall increase of reduction costs. Besides, large cities tended to have a cost disadvantage in terms of reducing carbon emissions. The heterogeneity in carbon reduction costs across regions and cities of

[2] https://www.cbsnews.com/news/the-most-polluted-places-on-earth/

different sizes also argues that carbon trading market is likely to be more efficient and flexible than command-and-control reduction regulations.

5.3.3 Determinants of Urban Carbon Reduction Costs

Upon estimating city-level carbon reduction costs, this study also analyses the determinants of carbon reduction costs. The panel regression approach with two-way fixed effects has been applied (Lewandowski 2017; Wang et al. 2017). Following Klepper and Peterson (2006), Wei et al. (2012), Du et al. (2015), and He et al. (2018), key factors that may influence carbon reduction costs have been considered, including carbon emissions intensity (*CEI*) (measured by the ratio of CO_2 emissions to GCP), energy mix (*EM*) (measured by the ratio of CO_2 emissions to energy use), industry mix (*IM*) (measured by the ratio of the output of the manufacturing industry to that of the service industry), capital (*K*) (measured by the ratio of capital to labour), city size (*CZ*) (measured by the natural logarithm of residential population) and trade openness (*TO*) (measured by the ratio of total value of foreign trade to GCP). The required data information is also drawn from *China City Statistical Yearbook* (NBSC 2003–2014a), *China Urban Construction Statistical Yearbook* (NBSC 2003–2014b), *Chinese Statistical Yearbook for Regional Economy* (NBSC 2003–2014c) and *China Statistical Yearbook* (NBSC 2003–2014d).

Table 5.1 provides the panel regression results. Only trade openness is not significant. The results are robust across models, and the AIC and SIC estimates argue that Model 5 fits best. Carbon reduction costs might be negatively influenced by the ratios of CO_2 emissions to GCP and CO_2 emissions to energy use might, but positively associated with the ratio of the output of the manufacturing industry to that of the service industry. Moreover, larger cities tended to have lower carbon reduction costs after controlling other impactors. The influence of capital was relatively slight. Specifically, the negative coefficient of carbon emissions intensity and the positive coefficient of its quadratic term imply that there was a U-shaped connection between urban carbon reduction cost and carbon emissions intensity. As China is tightening its carbon regulations, urban carbon reduction costs are likely to increase in the long-term.

5.4 Conclusions

This chapter employs a non-radial approach, the SBM-DEA method, to analyse city-level carbon reduction costs and potentials in China. A panel dataset including 286 PAA cities for the 2002–2013 period has been complied and analysed. Carbon reduction potentials are reflected by the estimated factor efficiency of carbon emissions. Carbon reduction costs are estimated through measuring shadow prices.

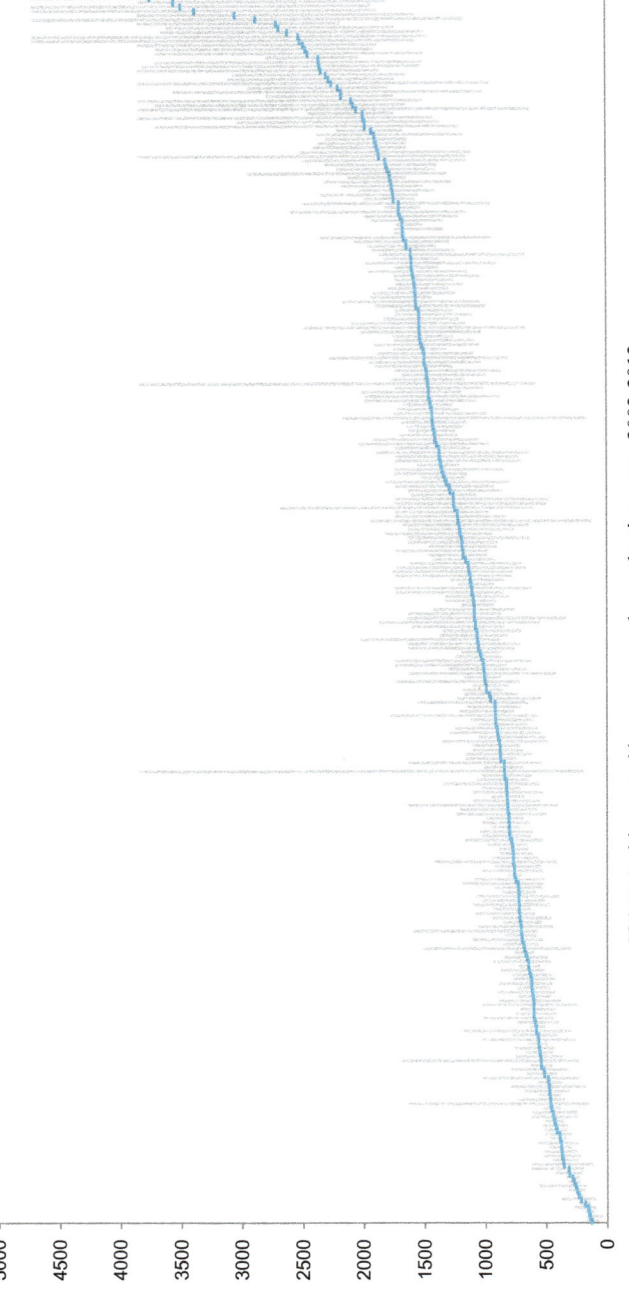

Fig. 5.3 Average, maximum and minimum urban carbon reduction costs, 2002–2013

Table 5.1 Regression results

Dependent Variable: Ln (carbon reduction cost)	Model 1	Model 2	Model 3	Model 4	Model 5	Model 6
CEI	−0.72**	−0.69**	−0.69**	−0.69**	−0.69**	−0.69**
	(0.02)	(0.01)	(0.01)	(0.01)	(0.01)	(0.01)
CEI Squared	0.04**	0.04**	0.04**	0.04**	0.04**	0.04**
	(0.00)	(0.00)	(0.00)	(0.00)	(0.00)	(0.00)
EM		−0.03**	−0.03**	−0.03**	−0.03**	−0.03**
		(0.00)	(0.00)	(0.00)	(0.00)	(0.00)
K			−0.00**	−0.00**	−0.00**	−0.00**
			(0.00)	(0.00)	(0.00)	(0.00)
CZ				−0.19**	−0.17**	−0.18**
				(0.05)	(0.05)	(0.05)
IM					0.03**	0.03**
					(0.01)	(0.01)
TO						−0.03
						(0.04)
Constant	7.70**	7.79**	7.81**	8.90**	8.78**	8.79**
	(0.02)	(0.02)	(0.02)	(0.30)	(0.30)	(0.30)
R^2	0.86	0.87	0.87	0.85	0.85	0.85
AIC	−2946	−3010	−3034	−3047	−3055	−3053
SIC	−2861	−2918	−2936	−2943	−2944	−2936

Note ** $p < 0.01$; Standard errors in parentheses

The results show that the factor efficiency of CO_2 emissions generally gained slow improvement for the 2002–2013 period, indicating that urban carbon reduction potential level showed a slightly decreasing trend. Nationally, the average and median values of urban carbon reduction costs were about 1,190 Yuan/tonne and 1,050 Yuan/tonne, respectively. The estimates reveal substantially geographical heterogeneities in carbon reduction costs and potentials across cities. In average, the eastern cities had smaller carbon reduction potential, while the northeastern cities had larger reduction potential. The eastern cities had higher carbon reduction cost whereas the northeastern cities had lower reduction cost. The carbon reduction cost of the eastern cities was slightly higher than those of the central and western cities. The annual average carbon reduction costs for the eastern, central, and western cities fluctuated during the studied period. However, the northeatern cities experienced a considerable increase of 94.9%. There was a U-shaped connection between urban carbon reduction cost and carbon emissions intensity. As China is tightening its carbon regulations, urban carbon reduction costs are likely to increase in the long-term. Overall, the heterogeneities in carbon reduction potentials and costs across cities argue that carbon trading market is likely to be more efficient and flexible than command-and-control reduction regulations.

Readers need bear in mind that the approach adopted in this study does not consider possible use of new technologies or technological heterogeneity across cities, and therefore the estimated reduction potential and cost results might be biased. Further research calls for more accurate techniques that include those ignored issues.

References

Almond, D., Chen, Y., Greenstone, M., & Li, H. (2009). Winter heating or clean air? Unintended impacts of China's Huai River policy. *American Economic Review*, 99(2), 184-90.

Aparicio, J., Ortiz, L., & Pastor, J. T. (2017). Measuring and decomposing profit inefficiency through the Slacks-Based Measure. *European Journal of Operational Research*, 260(2), 650-654.

Chang, W. Y., Wang, S., Song, X., & Zhong, F. (2022). Economic effects of command-and-control abatement policies under China's 2030 carbon emission goal. *Journal of Environmental Management*, 312, 114925.

Chang, Y. T., Zhang, N., Danao, D., & Zhang, N. (2013). Environmental efficiency analysis of transportation system in China: A non-radial DEA approach. *Energy Policy*, 58, 277-283.

Chen, X., Chen, G., Lin, M., Tang, K., & Ye, B. (2022). How does anti-corruption affect enterprise green innovation in China's energy-intensive industries?. *Environmental Geochemistry and Health*, 44, 2919–2942.

Chen, J., Gao, M., Cheng, S., Liu, X., Hou, W., Song, M., Li, D., & Fan, W. (2021). China's city-level carbon emissions during 1992–2017 based on the inter-calibration of nighttime light data. *Scientific Reports*, 11(1), 1-13.

Chen, J., Wang, L., & Li, Y. (2020). Research on the impact of multi-dimensional urbanization on China's carbon emissions under the background of COP21. *Journal of Environmental Management*, 273, 111123.

Chen, Q., Cai, B., Dhakal, S., Pei, S., Liu, C., Shi, X., & Hu, F. (2017). CO_2 emission data for Chinese cities. *Resources, Conservation and Recycling*, 126, 198-208.

Choi, Y., Zhang, N., & Zhou, P. (2012). Efficiency and abatement costs of energy-related CO_2 emissions in China: A slacks-based efficiency measure. *Applied Energy*, 98, 198-208.

Cui, J., Wang, C., Zhang, J., & Zheng, Y. (2021). The effectiveness of China's regional carbon market pilots in reducing firm emissions. *Proceedings of the National Academy of Sciences*, 118(52), e2109912118.

Dhakal, S. (2009). Urban energy use and carbon emissions from cities in China and policy implications. *Energy Policy*, 37(11), 4208-4219.

Du, Q., Wu, J., Cai, C., Li, Y., Zhou, J., & Yan, Y. (2021). Carbon mitigation by the construction industry in China: a perspective of efficiency and costs. *Environmental Science and Pollution Research*, 28(1), 314-325.

Glaeser, E. L., & Kahn, M. E. (2010). The greenness of cities: Carbon dioxide emissions and urban development. *Journal of Urban Economics*, 67(3), 404-418.

Gómez-Calvet, R., Conesa, D., Gómez-Calvet, A. R., & Tortosa-Ausina, E. (2014). Energy efficiency in the European Union: What can be learned from the joint application of directional distance functions and slacks-based measures?. *Applied Energy*, 132, 137-154.

Guo, X. D., Zhu, L., Fan, Y., & Xie, B. C. (2011). Evaluation of potential reductions in carbon emissions in Chinese provinces based on environmental DEA. *Energy Policy*, 39(5), 2352-2360.

Hailu, A., & Ma, C. (2017). The efficiency and distributional effects of China's carbon mitigation policies: A distance function analysis. UWA School of Agricultural and Environment Working Paper.

He, W., Wang, B., & Wang, Z. (2018). Will regional economic integration influence carbon dioxide marginal abatement costs? Evidence from Chinese panel data. Energy Economics, 74, 263-274

He, W., Zhang, B., & Ding, T. (2020). Sources of provincial carbon intensity reduction potential in China: A non-parametric fractional programming approach. *Science of The Total Environment*, 730, 139037.

Henderson, J. V. (2010). Cities and development. *Journal of Regional Science*, 50(1), 515-540.

IPCC. (2006). *Greenhouse Gas Inventory: IPCC Guidelines for National Greenhouse Gas Inventories*. United Kingdom Meteorological Office, Bracknell, England.

Klepper, G., & Peterson, S. (2006). Marginal abatement cost curves in general equilibrium: The influence of world energy prices. *Resource and Energy Economics*, 28(1), 1-23.

Kostka, G. (2016). Command without control: The case of China's environmental target system. *Regulation & Governance*, 10(1), 58-74.

Kuhn, L., Balezentis, T., Hou, L., & Wang, D. (2020). Technical and environmental efficiency of livestock farms in China: A slacks-based DEA approach. *China Economic Review*, 62, 101213.

Larsen, K., Pitt, H., Grant, M., & Houser, T. (2021). China's greenhouse gas emissions exceeded the developed world for the first time in 2019. Rhodium Group. https://rhg.com/research/chinas-emissions-surpass-developed-countries/#_ftn1

Lee, H. S. (2021). Slacks-based measures of efficiency and super-efficiency in presence of nonpositive data. *Omega*, 103, 102395.

Lewandowski, S. (2017). Corporate carbon and financial performance: The role of emission reductions. *Business Strategy and the Environment*, 26(8), 1196-1211.

Li, P., & Sun, W. (2018). Temporal evolution and influencing factors of energy consumption and related carbon emissions from the perspective of industrialization and urbanization in Shanghai, China. Sustainability, 10(9), 3054

Liao, Z., & Yao, Q. (2022). Flexibility is needed in China's national carbon market. *Nature Climate Change*, 12, 106-107.

Lin, B., & Moubarak, M. (2014). Mitigation potential of carbon dioxide emissions in the Chinese textile industry. *Applied Energy*, 113, 781-787.

Liu, X., Duan, Z., Shan, Y., Duan, H., Wang, S., & Song, J. (2019). Low-carbon developments in Northeast China: Evidence from cities. *Applied Energy*, 236, 1019-1033.

Lv, T., Hu, H., Zhang, X., Wang, L., & Fu, S. (2022). Impact of multidimensional urbanization on carbon emissions in an ecological civilization experimental area of China. *Physics and Chemistry of the Earth, Parts A/B/C*. DOI: https://doi.org/10.1016/j.pce.2022.103120

Meng, M., & Pang, T. (2022). Operational efficiency analysis of China's electric power industry using a dynamic network slack-based measure model. *Energy*, 123898.

Mohammadi, N., & Taylor, J. E. (2017). Urban energy flux: Spatiotemporal fluctuations of building energy consumption and human mobility-driven prediction. *Applied Energy*, 195, 810-818.

NBSC (National Bureau of Statistics of China) (2003–2014a). *China City Statistical Yearbook*. Beijing: China Statistic Press.

NBSC (2003–2014b). *China Urban Construction Statistical Yearbook*. Beijing: China Statistic Press.

NBSC (2003–2014c). *China Statistical Yearbook for Regional Economy*, Beijing: China Statistic Press.

NBSC (2003–2014d). *China Statistical Yearbook*. Beijing: China Statistic Press.

Shahbaz, M., Loganathan, N., Muzaffar, A. T., Ahmed, K., & Jabran, M. A. (2016). How urbanization affects CO_2 emissions in Malaysia? The application of STIRPAT model. *Renewable and Sustainable Energy Reviews*, 57, 83-93.

Shao, T., & Jin, H. (2020). A field investigation on the winter thermal comfort of residents in rural houses at different latitudes of northeast severe cold regions, China. Journal of Building Engineering, 32, 101476

Sueyoshi, T., Yuan, Y., & Goto, M. (2017). A literature study for DEA applied to energy and environment. Energy Economics, 62, 104-124

Tang, K., & Ma, C. (2022). The cost-effectiveness of agricultural greenhouse gas reduction under diverse carbon policies in China. *China Agricultural Economic Review*. https://doi.org/10.1108/CAER-01-2022-0008

Tang, K., He, C., Ma, C., & Wang, D. (2019). Does carbon farming provide a cost-effective option to mitigate GHG emissions? Evidence from China. *Australian Journal of Agricultural and Resource Economics*, 63(3), 575-592.

Tang, K., Yang, L., & Zhang, J. (2016). Estimating the regional total factor efficiency and pollutants' marginal abatement costs in China: A parametric approach. *Applied Energy*, 184, 230-240.

Tang, K., Liu, Y., Zhou, D., & Qiu, Y. (2021a). Urban carbon emission intensity under emission trading system in a developing economy: Evidence from 273 Chinese cities. *Environmental Science and Pollution Research*, 28(5), 5168-5179.

Tang, K., Wang, M., & Zhou, D. (2021b). Abatement potential and cost of agricultural greenhouse gases in Australian dryland farming system. *Environmental Science and Pollution Research*, 28(17), 21862-21873.

Tang, K., Xiong, C., Wang, Y., & Zhou, D. (2021c). Carbon emissions performance trend across Chinese cities: Evidence from efficiency and convergence evaluation. *Environmental Science and Pollution Research*, 28(2), 1533-1544.

Tang, K., Zhou, Y., Liang, X., & Zhou, D. (2021d). The effectiveness and heterogeneity of carbon emissions trading scheme in China. *Environmental Science and Pollution Research*, 28(14), 17306-17318.

Tone, K. (2001). A slacks-based measure of efficiency in data envelopment analysis. *European Journal of Operational Research*, 130(3), 498-509.

U.S. Energy Information Administration. (2018). *International Energy Outlook 2018*. US Energy Information Administration: Washington, DC, USA.

Vardanyan, M., & Noh, D. W. (2006). Approximating pollution abatement costs via alternative specifications of a multi-output production technology: A case of the US electric utility industry. *Journal of Environmental Management*, 80(2), 177-190.

Wang, J., Lv, K., Bian, Y., & Cheng, Y. (2017). Energy efficiency and marginal carbon dioxide emission abatement cost in urban China. *Energy Policy*, 105, 246-255.

Wang, Y., Chen, H., Long, R., Sun, Q., Jiang, S., & Liu, B. (2022). Has the sustainable development planning policy promoted the green transformation in China's resource-based cities?. *Resources, Conservation and Recycling*, 180, 106181.

Wei, C., Ni, J., & Du, L. (2012). Regional allocation of carbon dioxide abatement in China. *China Economic Review*, 23(3), 552-565.

Wu, Y. (2009). China's capital stock series by region and sector. The University of Western Australia Discussion Paper 09.02.

Wu, J., & Ma, C. (2019). The convergence of China's marginal abatement cost of CO_2: An emission-weighted continuous state space approach. *Environmental and Resource Economics*, 72(4), 1099-1119.

Wu, L., Chen, Y., & Feylizadeh, M. R. (2019). Study on the estimation, decomposition and application of China's provincial carbon marginal abatement costs. *Journal of Cleaner Production*, 207, 1007-1022.

Wu, J., Kang, Z. Y., & Zhang, N. (2017). Carbon emission reduction potentials under different polices in Chinese cities: A scenario-based analysis. *Journal of Cleaner Production*, 161, 1226-1236.

Wu, J., Ma, C., & Tang, K. (2019). The static and dynamic heterogeneity and determinants of marginal abatement cost of CO_2 emissions in Chinese cities. *Energy*, 178, 685-694.

Wu, J., Feng, Z., & Tang, K. (2021a). The dynamics and drivers of environmental performance in Chinese cities: A decomposition analysis. *Environmental Science and Pollution Research*, 28(24), 30626-30641.

Wu, J., Xu, H., & Tang, K. (2021b). Industrial agglomeration, CO_2 emissions and regional development programs: A decomposition analysis based on 286 Chinese cities. *Energy*, 225, 120239.

Wu, F., Huang, N., Liu, G., Niu, L., & Qiao, Z. (2020). Pathway optimization of China's carbon emission reduction and its provincial allocation under temperature control threshold. *Journal of Environmental Management*, 271, 111034.

Xu, L., Wang, J., Xiao, F., Sherif, E. B., & Awed, A. (2021). Potential strategies to mitigate the heat island impacts of highway pavement on megacities with considerations of energy uses. *Applied Energy*, 281, 116077.

Yang, L., Tang, K., Wang, Z., An, H., & Fang, W. (2017). Regional eco-efficiency and pollutants' marginal abatement costs in China: A parametric approach. *Journal of Cleaner Production*, 167, 619-629.

Yang, L., Yang, Y., Zhang, X., & Tang, K. (2018). Whether China's industrial sectors make efforts to reduce CO_2 emissions from production? A decomposed decoupling analysis. *Energy*, 160, 796-809.

Zarco-Soto, I. M., Zarco-Periñán, P. J., & Sánchez-Durán, R. (2021). Influence of cities population size on their energy consumption and CO_2 emissions: The case of Spain. *Environmental Science and Pollution Research*, 28(22), 28146-28167.

Zheng, S., Wang, R., Glaeser, E. L., & Kahn, M. E. (2011). The greenness of China: household carbon dioxide emissions and urban development. *Journal of Economic Geography*, 11(5), 761-792.

Zhou, H., Yang, Y., Chen, Y., & Zhu, J. (2018a). Data envelopment analysis application in sustainability: The origins, development and future directions. *European Journal of Operational Research*, 264(1), 1-16.

Zhou, P., Zhou, X., & Fan, L. W. (2014). On estimating shadow prices of undesirable outputs with efficiency models: A literature review. *Applied Energy*, 130, 799-806.

Zhou, Z., Liu, C., Zeng, X., Jiang, Y., & Liu, W. (2018b). Carbon emission performance evaluation and allocation in Chinese cities. *Journal of Cleaner Production*, 172, 1254-1272.

Zhu, Q., & Peng, X. (2012). The impacts of population change on carbon emissions in China during 1978–2008. *Environmental Impact Assessment Review*, 36, 1-8.

Chapter 6
Cost-Effectiveness of Agricultural Carbon Reduction in China

Kai Tang and Dong Wang

6.1 Introduction

The global communities nowadays widely support that it is vital to achieve carbon neutrality as early as possible. Growing scientific evidence argues that uncontrolled anthropogenic emissions of greenhouse gases (GHG), mainly carbon dioxide (CO_2) emissions associated with fossil fuels use, are likely to bring increasingly frequent fires, droughts and flooding, and 1.5 °C warming about 2030 (IPCC 2021). Therefore, deep and fast cuts in GHG emissions, which could stabilize rising temperatures and avoid the possible catastrophe, are in great demand. Accordingly, a growing number of countries have announced their carbon neutrality plans. In 2020, some major economies announced target dates for achieving carbon neutrality, many aiming for 2050 (e.g., Japan, Germany, and Canada).

China is the largest emitter of GHG in the world. Every year, China emits more GHG than the entire developed countries combined. In 2019, China emitted 27% of the world's GHG, while the US and India contributed 11% and 6%, respectively. Under both international and domestic pressures to do more to address global warming, China's leaders have recently pledged to peak its emissions within the next ten years, and become carbon neutral before 2060. This is the first long-term climate goal of the most-populous country, which could affect more than the country's 1.4 billion people. Considering China's ongoing rapid industrialization and urbanization, the goal is ambitious and challenging.

As one of the key sources of GHG emissions, agriculture contributes large amounts of CO_2, methane (CH_4) and nitrous oxide (N_2O) through, i.e., emissions from

K. Tang (✉)
School of Economics and Trade, Guangdong University of Foreign Studies, Guangzhou 510006, China
e-mail: francistang1988@hotmail.com

D. Wang
Crawford School of Public Policy, Australian National University, Acton, ACT 2601, Australia

fertiliser use or livestock. Totally, agriculture accounts for approximately 15 per cent of China's total GHG emissions, 90% of N_2O emissions, and 60% of CH_4 emissions, which is equivalent to about 2.7 times Australia's entire GHG emissions (Tang et al. 2016a, 2019; Yang et al. 2018; Tang and Ma 2022). In 2010, China's agriculture emitted 471 million tCO_2e CH_4 and 358 million tCO_2e N_2O (MEE 2018). In 2019, GHG emissions from crop and livestock production in China were about 0.8 billion tonnes CO_2e. Accordingly, agriculture is expected to make a great contribution to China's carbon–neutral progress.

Existing literature has demonstrated that that farming practices such as conservation tillage, crop stubble management, and afforestation are able to increase soil carbon and decrease carbon emissions (Havlík et al. 2012; Khataza et al. 2017; Tang et al. 2018). In addition, practices such as altering crop-pasture mix and enhancing livestock enterprise can also mitigate CH_4 and N_2O emissions (Bellarby et al. 2013; Herrero et al. 2016; Vasconcelos et al. 2018). Noticeably, adequate policy incentives are often requisite to inducing those practices (Thamo et al. 2017; Tang et al. 2019, 2021b).

Policymakers in China started to take measures to tackle climate change since the late 1990s. The most well-known agricultural policy is the Green for Grain program launched in 1999. It is one of the biggest conservation set-aside programs in the world. Another national effort was the eco-household program to boost renewable energy and biogas digesters use. However, these early policies related to carbon reduction in agricultural sector were usually outweighed by the impacts of pro-food policies that are motivated by food security targets (NDRC 2015; Tang et al. 2019; Yuan et al. 2021, 2022).

Since the 12th Five-Year Plan (FYP 2011–2015), the Chinese government started to implement more specific measures to decrease agricultural GHG emissions as a response to its ambitious national carbon reduction targets. Such measures include enhancing the formulation and implementation of relevant laws and regulations, adopting agricultural technologies that are more compatible with GHG reduction, promoting soil test-based fertilization practices to mitigate fertilizers applied and related N_2O emissions, extending straw mulching and no-tillage, feeding cattle with ammoniated straw, and transforming from traditional livestock herding to intensive livestock farming (NCCCC 2012). Among those measures, promoting soil test-based fertilization practices, an essential strategy reducing N_2O emissions, is the primary task.

Additionally, the government also set specific targets for agriculture. For example, the Ministry of Agriculture (MOA) ran schemes to mitigate carbon emissions through promoting agricultural production, i.e., increasing fertilizer utilization efficiency 1% annually and achieving 90 per cent utilization rate of soil test-based fertilization practices by 2020. In the 13th FYP (2016–2020), the Chinese government announced its first national agricultural GHG reduction targets: peaking agricultural N_2O emissions and reducing 18% carbon emissions per unit of GDP by 2020. These targets have been achieved mainly through direct financial incentives and command-and-control (CAC) regulations.

Recently, China has also run market-based reduction incentives considering the widely held belief that those market-based incentives could be more cost-effective than CAC regulations (Wu and Ma 2019; Tang et al. 2021a; Tang and Ma 2022). Neither the several emission trading pilot markets (ETPM) started in 2012, nor the national emission trading scheme (ETS) launched in 2021 includes agricultural GHG emissions. However, it is of policymakers' great interest to understand the cost-effectiveness of agricultural carbon reduction.

Despite the Chinese policymakers' interests in reducing agricultural GHG emissions, few has empirically assessed the cost-effectiveness of agricultural GHG reduction in the most-populous country. In this chapter, we address this topic by using a whole-farm-bioeconomic approach. Specifically, we explore the variations in crop-livestock enterprises, on-farm GHG emissions, and marginal abatement costs with various levels of market-based incentives. Given that livestock emit about 15% of global anthropogenic GHG emissions (Gerber et al. 2013; FAO 2013) and 50% of agricultural GHG emissions in China (Dong et al. 2008), the analysis includes livestock enterprise and focuses on cropping-livestock mixed farming. The estimated results may shed new light on the implications of China's agricultural reduction practices and policies.

6.2 Methodology

6.2.1 Study Region

This chapter focuses on the agriculture in the Loess Plateau, the 2nd largest arable land area in China and the largest rainfed agriculture area in East Asia. The plateau covers an area of 0.63 million km^2 in the upper-middle reaches of the Yellow River, consisting of tablelands, slopes and highly erodible hills (Fig. 6.1). More than 100 million people live on the plateau, resulting in a higher population density than the country's average.[1] Farming activities use about three-fourths of the plateau (Liu et al. 2003). Local rural communities produce about one-fifth live sheep in China. The plateau has a semi-arid continental monsoon climate with 200–600 mm annual precipitation. The long-term average value is about 450 mm. Most precipitation occurs in summer and early autumn (Nolan et al. 2008). Local soils, mainly characterized as loessial and sandy, have relatively low fertility due to heavy soil erosion and surface runoff (Wang et al. 2009).

The local agricultural system is characterized as smallholder mixed rainfed farming. Similar agricultural systems can also be found in south-eastern Kazakhstan, western Iran, and inland areas of South Africa. Most farmers have 0.5–3 ha land and run both crop and livestock enterprises, mainly self-replacing sheep flocks (SBNHAR 2011–2020). Farmland is used for cropping, including cereals and cash crops, and

---

[1] http://www.gov.cn/gzdt/att/att/site1/20110117/001e3741a2cc0e9e318c01.pdf.

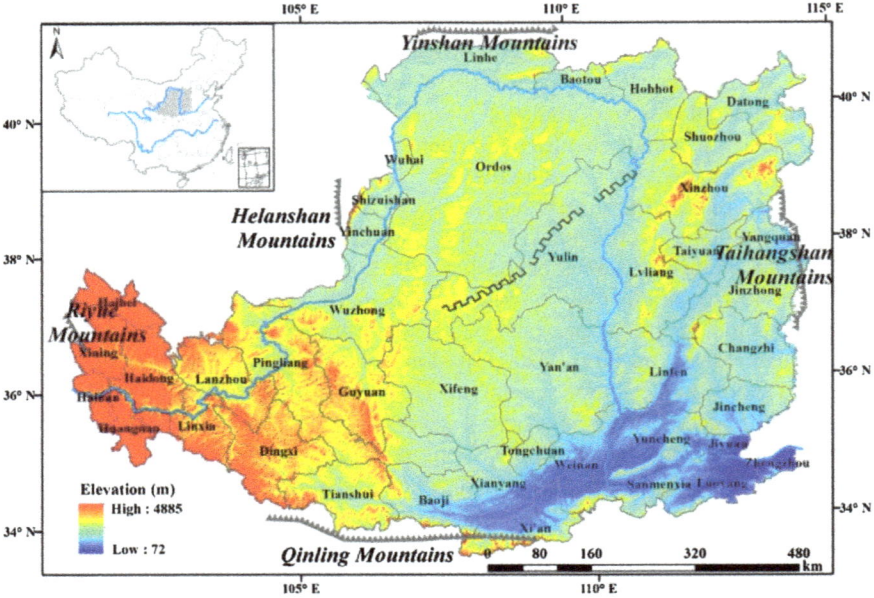

Fig. 6.1 The Loess Plateau, China. *Source* Wang et al. (2018)

pastures. Rainfall is usually the only water source for agricultural production (Liu et al. 2003). Because of water constraints, local crop yield is considerably lower (i.e., 50%) than the national average. The main crop is wheat (*Triticum aestivum*), which accounts for about one-third of crop production and one-fifth of farmland use (Nolan et al. 2008). Other crops include dry pea (*Pisum sativum*), oats (*Avena sativa*), maize (*Zea mays*), and rapeseed (*Brassica napus*).

6.2.2 Whole-Farm-Bioeconomic Approach

This chapter uses a whole-farm-bioeconomic approach in Tang et al. (2019). The approach is an updated version of the ones in Hailu et al. (2011) and Tang et al. (2018). The approach mainly includes a dynamic land use optimisation model. The model aims to maximize farm's net gross margin, which represents the optimal land use sequence choice affecting by farming running history. The constrains include environmental, managerial, and financial factors of the represented smallholder rainfed mixed farming. The results reveal the optimal enterprise, including optimal land use allocation, of the farm.

The approach calculates on-farm GHG emissions based on a revised Intergovernmental Panel on Climate Change inventory method (Tang et al. 2019). All emissions have been converted into carbon dioxide equivalents (CO_2e) using global warming potential equivalence values. The market-based incentives are formed as a flat rate emission tax. Each ton of GHG emitted should be taxed with a fixed amount. 11 emission tax scenarios are included, ranging from 0 to 500 Chinese Yuan (¥)/tCO2e in increments of ¥50/tCO2e. Similar scenarios can be found in some recent related studies (Thamo et al. 2017; Tang et al. 2018, 2019; Tang and Hailu 2020). These scenarios are both consistent with the existing literature and broad enough to show the uncertainty in policies, increasing estimates' robustness.

The chapter sets the annual precipitation as 450 mm, reflecting the mean level in the studied region. Cropping enterprise consists of wheat, oats, rapeseed, and dry pea. Farmland can also be used for leguminous pastures, which provide food for sheep the representative livestock. Both cropping and livestock enterprises generate profits. The approach uses the 2015 farm-gate prices. For more details about parameters and equations used, we refer the readers to Tang et al. (2019) and Tang and Hailu (2020).

6.3   Results

In the analysis, the optimization horizon is a decade. In this section, the results shown are the annual average of the optimization horizon for a representative smallholder mixed farm.

6.3.1   Base Scenario

In the base scenario (emission tax rate is 0), annual net gross margin brought by the optimal enterprises is ¥3545/ha. Farmland is approximately equally divided between cropping and livestock enterprises. Wheat and oats are the main crops grown. About 20% of farmland is allocated for wheat, while 21% of that is used for oats. 9% of farmland is devoted to rapeseed. Nearly all the remainder is covered by leguminous pastures since share of land for dry pea is less than 1%. Table 6.1 show that the popular rotations in the optimal enterprises consist of continuous pastures, pastures-oats, pastures-wheat-oats, and pastures-rapeseed-wheat rotations (Table 6.1). Noticeably, the choice of those rotations is affected by differing soil characteristics. The simulated annual yields of crops are around 3t/ha. Those estimated are broadly in line with the related literature on the studied plateau (Tsunekawa et al. 2014; Tang et al. 2019; Tang and Hailu 2020).

The base scenario is accompanied by an annual on-farm emissions rate of $2tCO_2e$/ha. The majority of GHG emitted, more than 80% in this case, are from livestock enterprise. Nitrogen-fixing pastures emit around 13% of the total emissions. The reminder is from fertilization application and crops residues. Scientific evidence has shown that GHG emissions from livestock enterprise are considerable, especially in similar rainfed farming systems (Thamo et al. 2017; Tang and Hailu 2020).

Table 6.1 Optimal rotations by soils with differing emission tax rates

Emission tax rate (¥/t CO2e)	Optimal rotations			
	Sandy soils	Loess sands	Loessial soils	Bedrock
0	POO, PRO, PPOD	POO, POP, POW, PRO, PWO, PRW, WPRW	PPP, POP, PPO, PWP, PPW, PRW	PRW, PRWP
50	POW, PWW, PRW,	PPP, POO, POP, PRW, WPRW	POP, POW, PWO, PWW, PWPW, PWD, PRW, DOP, OODR	POO, POOD
100	PPP, PRW, PDR	PWW, PPW, PRW, PDW, WPRW	PWW, WPW, WPP, PRW, PDW, PPD, WPD	PPP, WDP
150	PPO, WPWO, PPO	PRW, PDW, PWD, DRW	PWW, PDW, PWD, PDWD	PWW, PRW, PRWD
200	PWW, PDW, DPDW	PWW, PRW, PDW, WPDW	PWW, PRW, PDW, PWD, PDWD, DRW, DPDW	PPO, PDW
250	PRW, PDR, DRW	PWW, PRW, PDW, PWD	PWW, PRW, WPRW, PDW, PWD	PWW, PDW, PDWD
300	PWW, PWD, PDW	PWW, PDW, PDWD	PWW, PDW, PWD, PDR, WPDO	PWW, DRW
350	PWW, PDW, DPDW, WPR	PWW, PDW, PWD, DPDW	PWW, PDW, PWD, DPDW, DRW	DRW, DRWD
400	PDW, DRW, DCWR	PDW, PDWD, DRW	PWW, PDW, PWD, PDWD, DPDW	PDR, DRW
450	PDW, PWD, PDR, DRW	PWW, PDW, PWD, PDWD, DRW	PWW, PDW, PWD, PDWD, DPDW, RDR	PDW, PDWD
500	PDW, PDWD, WDR	PWW, PDW, PWD, DPDW, WDR	PWW, PDW, PWD, PDWD, DPDW, DRD	PWW, PDW, PDWD

Note W = wheat (*Triticum aestivum*); O = oats (*Avena sativa*); R = rapeseed (*Brassica napus*); D = dry pea (*Pisum sativum*); P = leguminous pastures

6.3.2 Various Emission Tax Scenarios

The chapter illustrates the optimal net gross margin and the related on-farm GHG emissions in Fig. 6.2 and the farmland mixes for the optimal enterprises with differing emission tax rates in Fig. 6.3. When the emission tax rate is set as ¥50/tCO2e, the net gross margin is ¥3460/ha, and the emissions are 8 per cent lower than the base scenario. More land is devoted to growing wheat and dry pea, while less to producing pastures, oats, and rapeseed. A ¥100/tCO2e emission tax rate results in ¥3410/ha net gross margin and further decrease in on-farm emissions, which are 17% lower compared with the base value.

Furthermore, if the tax rate is ¥150/tCO2e, then the net gross margin reduces ¥145/ha and the emissions decrease by about 30%, compared with the base scenario. Farmers continue expanding wheat and dry pea enterprises and reducing pastures grown. Pastures-wheat and pastures-dry-pea-wheat rotations become popular (Table 6.1). When the rate becomes higher, farmers tend to proportionately decrease emissions from cropping and livestock enterprises (Fig. 6.4). The wheat yield remains stable (3.25 t/ha) in most wheat-included rotations, however, could be increased by about 12% in pastures-dry-pea-wheat rotations.

When the emission tax rate is higher than ¥150/tCO2e, increasing emission tax rate is accompanied by loss of net gross margin. It should be noted that on-farm emissions reduce at a considerably slow rate with higher tax rate. Wheat and dry pea become the predominant crops in the optimal enterprises. Farmland devoted to

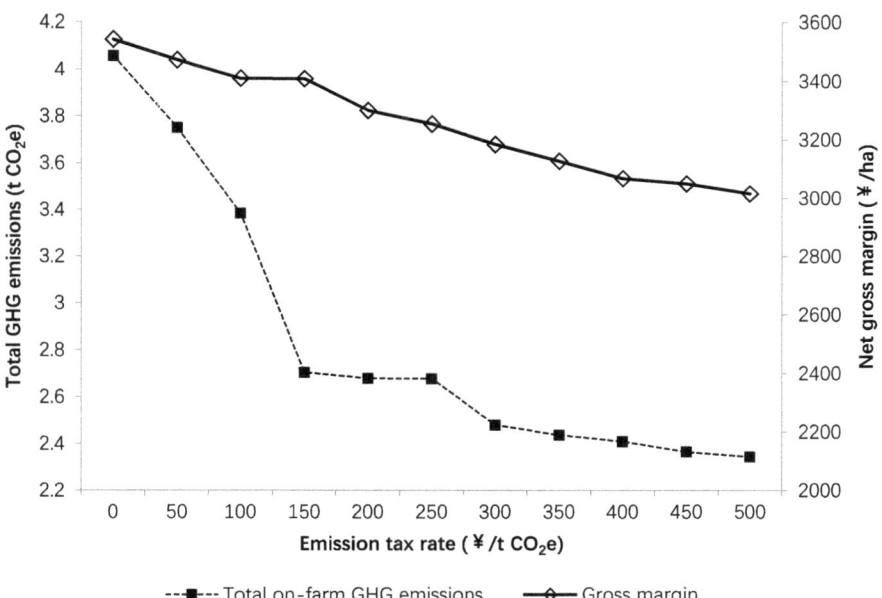

Fig. 6.2 Optimal net gross margin and on-farm GHG emissions with differing emission tax rates

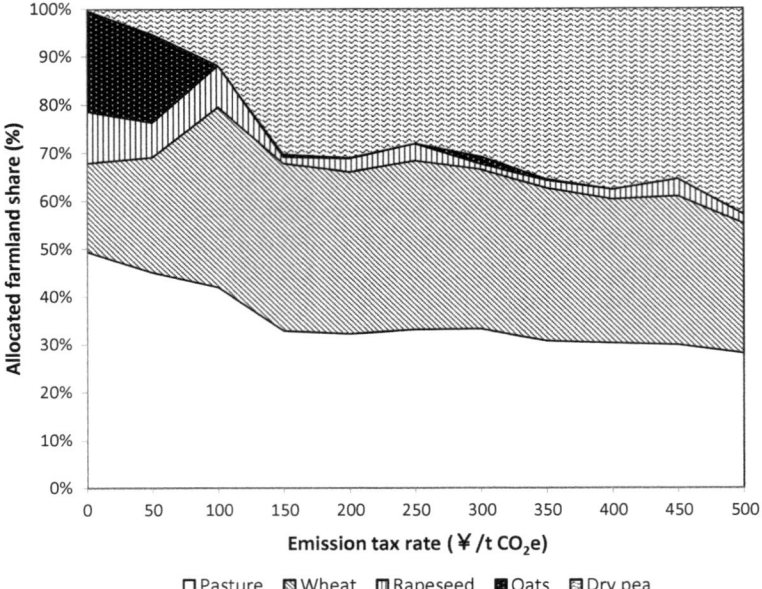

Fig. 6.3 Optimal farmland structures with differing emission tax rates

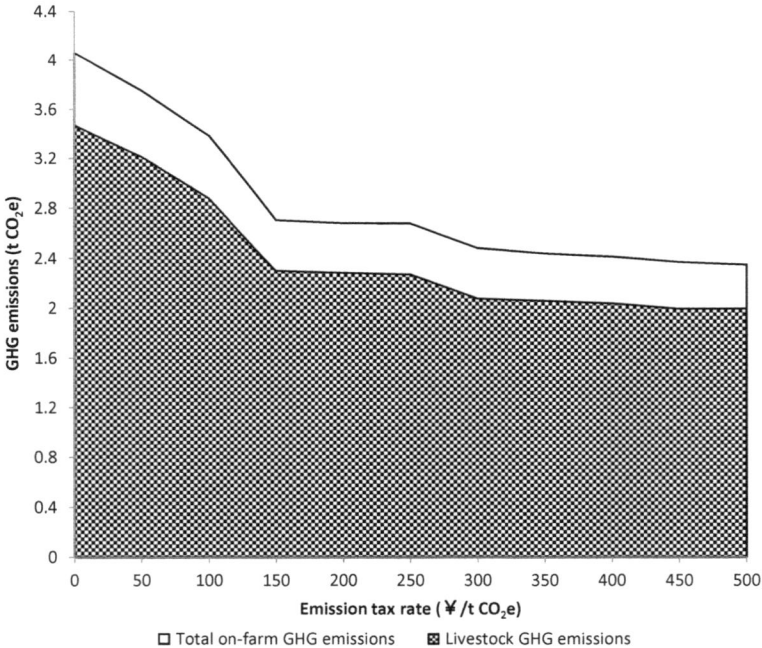

Fig. 6.4 On-farm emissions with differing emission tax rates

pastures remains stable roughly. The frequently chosen rotations are pastures-wheat and pastures-dry-pea-wheat. If the tax rate is set as ¥500/tCO2e, then the net gross margin reduces ¥540/ha and the emissions decrease by about 40%, compared with the base scenario. Dry pea, wheat and pastures grown are the dominating farmland use types. A tiny share of farmland is used for rapeseed cultivation.

6.4 Discussion

6.4.1 Cropping-Livestock Shifting

The analysis implies that high emission tax rate induces crop-dominating enterprises in the mixed farming studied. Globally, GHG emissions from livestock are on par with those from transportation sector (IPCC 2014; Reisinger and Clark 2018). Scientific evidence has shown that cropping enterprise is much less emission-intensive than livestock (Fiala 2008; Herrero et al. 2016; Tang et al. 2016b; Hawkins et al. 2018), i.e., cropping contributes to less than 20 per cent on-farm emissions in our case. Averagely, the annual emissions per ha for livestock enterprise are about six times the value of cropping enterprise. High emission tax rate indicates high production cost for emission-intensive enterprise. Rationally, farmers should respond by reducing livestock enterprise and increasing cropping, represented by lower share of farmland devoted to pastures grown in our analysis, to reduce costs and maintain profits.

Shifting from meat-based diets to plant-based diets has been identified as a key opportunity for tacking climate change (Schiermeier 2019; Kim et al. 2020). However, in many low- and middle-income countries in Asia and central and South America including China, the diet is continuing shifting more towards meat as disposable incomes increase (Fig. 6.5). People's appetite for meat is still growing. In China, the per capita consumption of certain meats increases rapidly, but still much lower than the levels in developed economies. China's annual per capita consumption of beef, for example, jumped from 4.88 kg in 2016 to 5.95 kg in 2019, while the number was more than 26 kg in the US in 2019 (OECD 2021). This highlights the substantial room for future increase as middle classes in China continue expanding.

The question is then: how could we find a balance between mitigating GHG emissions and increasing meat supply? A key action to resolve this dilemma is to innovate in livestock sector. Employing autonomous monitoring of livestock health and improving feed can play essential roles. Enhancing the genetic potential of livestock, their reproduction, health and liveweight gain rates are useful methods for abating GHG emissions per unit of product (Herrero et al. 2016; Grossi et al. 2019; Reisinger et al. 2021). Technical and management interventions, including using feed additives, improving feed digestibility and managing manure could also contribute to the resolution of the trade-offs between increasing livestock products supply and reducing agricultural GHG emissions (Adegbeye et al. 2019; Elghandour et al. 2020; Honan et al. 2021). The supply–demand gap could also create opportunities for

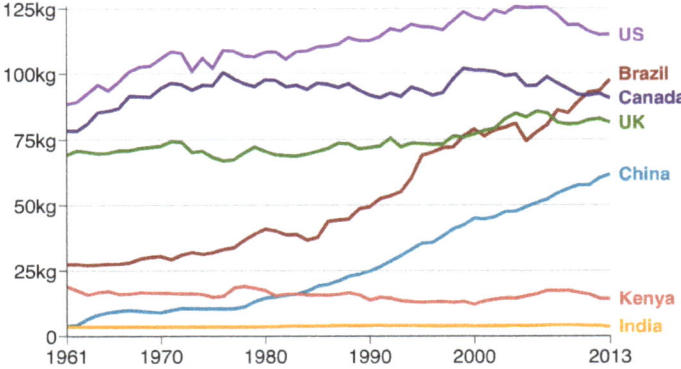

Fig. 6.5 Average annual meat consumption per capita in some countries. *Source* FAO/Our World in Data

international suppliers with more sustainable livestock sector including Australia, New Zealand and the US.

6.4.2 More Dry Pea Rotations

The analysis also indicates that cropping-livestock mixed farmers are likely to use more dry-pea-included rotations as an optimal response to increasing emission tax rate. For cropping enterprise, the combined effect of anthropogenic nitrogenous fertilizer applied and nitrogenous fixation crops is the key cause of N_2O emissions (Zhang et al. 2013; Smith 2017). Nitrogenous fixation crops, such as dry pea, contain symbiotic bacteria in their roots and produce nitrogen helping the plant growing need (Gan et al. 2009; Barton et al. 2011). Such a biophysical nitrogenous fixation process provides a natural way to supply nitrogen for nitrogenous fixation crops and subsequent crops, thus reducing the amount of anthropogenic nitrogenous fertilizer needed and the associated GHG emissions. Scientific evidence has shown that biophysical nitrogenous fixation makes dry-pea-included rotations smaller emission source compared with oats-based and rapeseed-based rotations in many rainfed agricultural regions (Asgedom and Kebreab 2011; Rajaniemi et al. 2011; Brock et al. 2016).

Furthermore, the results also show that the wheat yield could be increased in pastures-dry-pea-wheat rotations when wheat is grown as the subsequent crop. Such an effect might be further strengthened by enhancing the synchrony between nitrogen mineralisation from dry pea residues and the peak nitrogen demand of the subsequent crop (Tang et al. 2018). Overall, dry-pea-included rotations become popular when farmers have to pay high emission tax.

6.4.3   Marginal Abatement Costs

Results in Fig. 6.2 say that moderate emission tax rates are expected to result in considerable cuts in on-farm emissions and the accompanied loss of net gross margin may be small. A reduction of 8% emissions can be achieved by a rate of ¥50/tCO2e, accompanied by 2.4% net gross margin loss. Reduction of 17 and 30% emissions are achieved with ¥100/tCO2e and ¥150/tCO2e, respectively. The associated net gross margin loss can be 3.8% and 4.1%, respectively. The above estimates reveal that the marginal abatement costs of 8, 17 and 30% on-farm emissions could be less than ¥50/tCO2e, ¥100/tCO2e and ¥150/tCO2e, respectively.

China has run several regional pilot emission trading markets since 2013 and officially launched its national carbon market on July 2021. Observed carbon prices under the pilots and the recent national carbon market usually fluctuated between ¥50/tCO2e and ¥100/tCO2e. Considerable reduction might be achieved in agriculture if it could be provided incentives at similar levels. Reducing agricultural carbon emissions are cost-effective in China.

Moreover, it should be noted that the current national carbon market only includes 2225 coal- and gas-fired electricity plants. The carbon price is still considerably low since the initial allowances allocation are thought to be too loose and many key energy consumers such as metallurgical industry are not involved (Nogrady 2021).[2] It is expected that the national market will cover more industries and issue more reasonable allowances allocation, which might boost the price. Therefore, if agriculture could be provided incentives at comparable levels, the emissions reduced might be more substantial.

6.5   Conclusions

This chapter addresses the cost-effectiveness of agricultural GHG reduction in China by using a whole-farm-bioeconomic approach. Specifically, the analysis, which includes livestock enterprise and focuses on cropping-livestock mixed farming, explores the variations in crop-livestock enterprises, on-farm GHG emissions, and marginal abatement costs with various levels of market-based incentives.

The analysis implies that high emission tax rate induces crop-dominating enterprises in the mixed farming studied. Cropping-livestock mixed farmers are likely to use more dry-pea-included rotations as an optimal response to increasing emission tax rate. Moreover, moderate emission tax rates are expected to result in considerable cuts in on-farm emissions and the accompanied loss of net gross margin may be small. The estimates reveal that the marginal abatement costs of 8, 17 and 30% on-farm emissions could be less than ¥50/tCO2e, ¥100/tCO2e and ¥150/tCO2e, respectively.

---

[2] https://fortune.com/2021/07/16/china-carbon-emissions-trading-market-reform/.

Observed carbon prices under the pilots and the recent national carbon market imply that reducing agricultural carbon emissions are cost-effective in China. Considerable reduction might be achieved in agriculture if it could be provided incentives at similar levels. Farmers might be involved in the national carbon market since they would gain additional profits through selling credits to those with higher marginal abatement costs, thus promoting the profitability and contributing to narrowing urban–rural income gap. The estimated results may shed new light on the implications of China's agricultural reduction practices and policies.

References

Adegbeye, M. J., Elghandour, M. M., Monroy, J. C., Abegunde, T. O., Salem, A. Z., Barbabosa-Pliego, A., & Faniyi, T. O. (2019). Potential influence of Yucca extract as feed additive on greenhouse gases emission for a cleaner livestock and aquaculture farming-A review. *Journal of Cleaner Production*, 239, 118074.

Asgedom, H., & Kebreab, E. (2011). Beneficial management practices and mitigation of greenhouse gas emissions in the agriculture of the Canadian Prairie: A review. *Agronomy for Sustainable Development*, 31(3), 433-451.

Barton, L., Butterbach-Bahl, K., Kiese, R. and Murphy, D.V. (2011). Nitrous oxide fluxes from a grain–legume crop (narrow-leafed lupin) grown in a semiarid climate, *Global Change Biology* 17, 1153-1166.

Bellarby, J., Tirado, R., Leip, A., Weiss, F., Lesschen, J. P., & Smith, P. (2013). Livestock greenhouse gas emissions and mitigation potential in Europe. *Global Change Biology*, 19(1), 3-18.

Brock, P. M., Muir, S., Herridge, D. F., & Simmons, A. (2016). Cradle-to-farmgate greenhouse gas emissions for 2-year wheat monoculture and break crop-wheat sequences in south-eastern Australia. *Crop and Pasture Science*, 67(8), 812-822.

Dong, H., Li, Y., Tao, X., Li, N., & Zhu, Z. (2008). China's greenhouse gas emissions from agricultural activities and its mitigation strategy. *Transactions of the CSAE*, 24 (10), 269-273.

Elghandour, M. M. Y., Tan, Z. L., Abu Hafsa, S. H., Adegbeye, M. J., Greiner, R., Ugbogu, E. A., Cedillo Monroy, J., & Salem, A. Z. M. (2020). Saccharomyces cerevisiae as a probiotic feed additive to non and pseudoruminant feeding: A review. *Journal of Applied Microbiology*, 128(3), 658-674.

FAO (The Food and Agriculture Organization of The United Nations) (2013). Key facts and findings. http://www.fao.org/news/story/en/item/197623/icode/

Fiala, N. (2008). Meeting the demand: An estimation of potential future greenhouse gas emissions from meat production. *Ecological Economics*, 67(3), 412-419.

Gan, Y. T., Campbell, C. A., Jansen, H. H., Lemke, R., Liu, L. P., Basnyat, P., & McDonald, C. L. (2009). Carbon input to soil by oilseed and pulse crops in semiarid environment. *Agriculture, Ecosystem & Environment*, 132, 290-297.

Gerber, P. J., Steinfeld, H., Henderson, B., Mottet, A., Opio, C., Dijkman, J., Falcucci, A. & Tempio, G. (2013). Tackling climate change through livestock: A global assessment of emissions and mitigation opportunities. Food and Agriculture Organization of the United Nations, Rome.

Grossi, G., Goglio, P., Vitali, A., & Williams, A. G. (2019). Livestock and climate change: impact of livestock on climate and mitigation strategies. *Animal Frontiers*, 9(1), 69-76.

Hailu, A., Durkin, J., Sadler, R., & Nordblom, T. L. (2011). Agent-based modelling study of shadow, saline water table management in the Katanning catchment, Western Australia. Research Report for RIRDC Project No. PRJ-000578.

Havlík, P., Valin, H., Mosnier, A., et al. (2012). Crop productivity and the global livestock sector: Implications for land use change and greenhouse gas emissions. *American Journal of Agricultural Economics*, 95(2), 442-448.

Hawkins, J., Ma, C., Schilizzi, S., & Zhang, F. (2018). China's changing diet and its impacts on greenhouse gas emissions: an index decomposition analysis. *Australian Journal of Agricultural and Resource Economics*, 62(1), 45-64.

Herrero, M., Henderson, B., Havlík, P., et al. (2016). Greenhouse gas mitigation potentials in the livestock sector. *Nature Climate Change*, 6(5), 452-461.

Honan, M., Feng, X., Tricarico, J. M., & Kebreab, E. (2021). Feed additives as a strategic approach to reduce enteric methane production in cattle: modes of action, effectiveness and safety. *Animal Production Science*. https://doi.org/10.1071/AN20295

IPCC (2014). *Climate Change 2014: Mitigation of Climate Change. Contribution of Working Group III to the Fifth Assessment Report of the Intergovernmental Panel on Climate Change*. Cambridge University Press, Cambridge, United Kingdom and New York, NY, USA.

IPCC (2021). *Climate Change 2021: The Physical Science Basis. Contribution of Working Group I to the Sixth Assessment Report of the Intergovernmental Panel on Climate Change*. Cambridge University Press. In Press. https://www.ipcc.ch/report/ar6/wg1/downloads/report/IPCC_AR6_WGI_Full_Report.pdf

Khataza, R. R., Hailu, A., Kragt, M. E., & Doole, G. J. (2017). Estimating shadow price for symbiotic nitrogen and technical efficiency for legume-based conservation agriculture in Malawi. *Australian Journal of Agricultural and Resource Economics*, 61(3), 462-480.

Kim, B. F., Santo, R. E., Scatterday, A. P., Fry, J. P., Synk, C. M., Cebron, S. R., Mekonnen, M.M., Hoekstra, A.Y., De Pee, S., Bloem, M.W., & Nachman, K. E. (2020). Country-specific dietary shifts to mitigate climate and water crises. *Global Environmental Change*, 62, 101926.

Liu, J., Liu, M., Zhuang, D., Zhang, Z., & Deng, X. (2003). Study on spatial pattern of land-use change in China during 1995–2000. *Science in China Series D: Earth Sciences*, 46(4), 373-384.

MEE (Ministry of Ecological Environment) (2018). Third National Communication on Climate Change of the People's Republic of China. https://unfccc.int/sites/default/files/resource/China_NC3_Chinese_0.pdf

NCCCC (National Coordination Committee on Climate Change) (2012). Second National Communication on Climate Change of the People's Republic of China. China Planning Press, Beijing.

NDRC (National Development and Reform Commission) (2015). Intended Nationally Determined Contribution (INDC). Available at: http://www4.unfccc.int/submissions/INDC/Published%20Documents/China/1/China's%20INDC%20-%20on%2030%20June%202015.pdf

Nogrady, B. (2021). China launches world's largest carbon market: But is it ambitious enough?. *Nature*, 595(7869), 637-637.

Nolan, S., Unkovich, M., Yuying, S., Lingling, L., & Bellotti, W. (2008). Farming systems of the Loess Plateau, Gansu Province, China. *Agriculture, Ecosystems & Environment*, 124(1), 13-23.

OECD (2021). Meat consumption (indicator). DOI: https://doi.org/10.1787/fa290fd0-en

Rajaniemi, M., Mikkola, H., & Ahokas, J. (2011). Greenhouse gas emissions from oats, barley, wheat and rye production. *Agronomy Research*, 9(1), 189-195.

Reisinger, A., & Clark, H. (2018). How much do direct livestock emissions actually contribute to global warming?. *Global Change Biology*, 24(4), 1749-1761.

Reisinger, A., Clark, H., Cowie, A. L., Emmet-Booth, J., Gonzalez Fischer, C., Herrero, M., Howden, M., & Leahy, S. (2021). How necessary and feasible are reductions of methane emissions from livestock to support stringent temperature goals?. *Philosophical Transactions of the Royal Society A*, 379(2210), 20200452.

SBNHAR (Statistical Bureau of the Ningxia Hui Autonomous Region) (2011–2020). Ningxia Statistical Yearbook. China Statistics Press, Beijing.

Schiermeier, Q. (2019). Eat less meat: UN climate-change report calls for change to human diet. *Nature*, 572(7769), 291-293.

Smith, K. A. (2017). Changing views of nitrous oxide emissions from agricultural soil: Key controlling processes and assessment at different spatial scales. *European Journal of Soil Science*, *68*(2), 137–155.

Tang, K., Hailu, A., Kragt, M. E., & Ma, C. (2016a). Marginal abatement costs of greenhouse gas emissions: Broadacre farming in the Great Southern Region of Western Australia. *Australian Journal of Agricultural and Resource Economics*, *60*, 459–475.

Tang, K., He, C., Ma, C., & Wang, D. (2019). Does carbon farming provide a cost-effective option to mitigate GHG emissions? Evidence from China. *Australian Journal of Agricultural and Resource Economics*, 63(3), 575-592.

Tang, K., & Hailu, A. (2020). Smallholder farms' adaptation to the impacts of climate change: Evidence from China's Loess Plateau. *Land Use Policy*, 91, 104353.

Tang, K., Hailu, A., Kragt, M.E., & Ma, C. (2018). The response of broadacre mixed crop-livestock farmers to agricultural greenhouse gas abatement incentives. *Agricultural Systems*, 160, 11-20.

Tang, K., & Ma, C. (2022). The cost-effectiveness of agricultural greenhouse gas reduction under diverse carbon policies in China. *China Agricultural Economic Review*. https://doi.org/10.1108/CAER-01-2022-0008

Tang, K., Kragt, M. E., Hailu, A., & Ma, C. (2016b). Carbon farming economics: What have we learned?. *Journal of Environmental Management*, 172, 49-57.

Tang, K., Liu, Y., Zhou, D., & Qiu, Y. (2021a). Urban carbon emission intensity under emission trading system in a developing economy: Evidence from 273 Chinese cities. *Environmental Science and Pollution Research*, 28(5), 5168–5179.

Tang, K., Wang, M., & Zhou, D. (2021b). Abatement potential and cost of agricultural greenhouse gases in Australian dryland farming system. *Environmental Science and Pollution Research*, 28(17), 21862–21873.

Thamo, T., Addai, D., Pannell, D. J., Robertson, M. J., Thomas, D. T., & Young, J. M. (2017). Climate change impacts and farm-level adaptation: Economic analysis of a mixed cropping-livestock system. *Agricultural Systems*, 150, 99-108.

Vasconcelos, K., Farinha, M., Bernardo, L., et al. (2018). Livestock-derived greenhouse gas emissions in a diversified grazing system in the endangered Pampa biome, Southern Brazil. *Land Use Policy*, 75, 442-448.

Wang, S., Fu, B., Chen, H., & Liu, Y. (2018). Regional development boundary of China's Loess Plateau: Water limit and land shortage. *Land Use Policy*, 74, 130-136.

Wang, Y., Zhang, X., & Huang, C. (2009). Spatial variability of soil total nitrogen and soil total phosphorus under different land uses in a small watershed on the Loess Plateau, China. *Geoderma*, 150(1-2), 141-149

Wu, J., & Ma, C. (2019). The convergence of China's marginal abatement cost of CO_2: An emission-weighted continuous state space approach. *Environmental and Resource Economics*, 72(4), 1099-1119.

Yang, L., Yang, Y., Zhang, X., & Tang, K. (2018). Whether China's industrial sectors make efforts to reduce CO_2 emissions from production? A decomposed decoupling analysis. *Energy*, 160, 796-809.

Yuan, F., Tang, K., & Shi, Q. (2021). Does Internet use reduce chemical fertilizer use? Evidence from rural households in China. *Environmental Science and Pollution Research*, 28(5), 6005–6017.

Yuan, F., Tang, K., Shi, Q., Qiu, W., & Wang, M. (2022). Rural women and chemical fertiliser use in rural China. *Journal of Cleaner Production*, 344, 130959.

Zhang, W. F., Dou, Z. X., He, P., et al. (2013). New technologies reduce greenhouse gas emissions from nitrogenous fertilizer in China. *Proceedings of the National Academy of Sciences*, 110(21), 8375-8380.

Chapter 7
Investigating the Impact of Carbon Emission Trading on Urban Carbon Emissions in China

Kai Tang and Yichun Liu

7.1   Introduction

At the 2015 United Nations Climate Change Conference (COP 21) in Paris in December 2015, 195 countries adopted the first-ever legally binding global climate change agreement, known as the Paris Agreement. The Agreement aims at limiting global warming to well below 2 °C, preferably to 1.5 °C, compared to pre-industrial levels.[1] To achieve this ambitious goal, it is widely believed that the global community needs to peak global greenhouse gas emissions soon and to reach net-zero emissions by the middle of the twenty-first century (Meinshausen et al. 2009; Sachs et al. 2016; van Soest et al. 2021).[2] So there needs to be a transition from the current socioeconomic system that uses enormous amounts of fossil fuels and has high carbon emissions to a system that uses clean energy that produces limited emissions (Rogelj et al. 2016; Peake and Ekins 2017; Fankhauser et al. 2022). However, such a transition will not happen automatically; it requires policy instruments, such as command-and-control and market-based instruments (Michaelowa et al. 2018; Roberts et al. 2018; Andresen et al. 2021; Tang et al. 2020a; Tang and Ma 2022).

Emissions trading system (ETS) is a market-based instrument, which influences actors' behaviour by changing their economic incentive structure and internalises the external costs of pollutants, such as greenhouse gas emissions (Bullock 2012;

---

[1] https://unfccc.int/sites/default/files/english_paris_agreement.pdf.

[2] https://unfccc.int/process-and-meetings/the-paris-agreement/the-paris-agreement#:~:text=The%20Paris%20Agreement%20is%20a,compared%20to%20pre%2Dindustrial%20levels.

K. Tang (✉)
School of Economics and Trade, Guangdong University of Foreign Studies, Guangzhou 510006, China
e-mail: francistang1988@hotmail.com

Y. Liu
Department of Accounting and Finance, Lancaster University, Lancaster LA1 4YW, UK

Meleo et al. 2016; Tang et al. 2021a, c). Under an ETS, a cap is established either on emissions from all greenhouse gases or just some, such as CO_2, within a compliance period. Emission allowances, which are treated as a commodity and can be traded on the market, are then provided according to the cap level (Aldy and Stavins 2012). Those allowances can be allocated for free or through an auction. Theoretically, emissions reductions could be achieved at least cost through a highly effective ETS. Compared with command-and-control instruments, ETS is thought to be a more efficient tool to reduce greenhouse gas emissions due to its potential advantages such as cost-effectiveness (Bakam et al. 2012; Ranson and Stavins 2016; Schneider et al. 2017) and flexibility (Miola et al. 2011; Narassimhan et al. 2018). Besides, a well-designed ETS might induce innovation (Rogge et al. 2011; Borghesi et al. 2015; Verbruggen et al. 2019; Yao et al. 2021) and stimulate productivity (Zhou et al. 2020; Wu and Wang 2022).

Since the middle of the 2000s, several emissions trading systems have been launched. The EU ETS, set up in 2005, is the first international ETS globally. Besides the EU ETS, national or sub-national systems are already operating or under development in Canada (Québec Cap-and-Trade), China, Japan (Japan Voluntary Emission Trading Scheme (JVETS)), New Zealand (New Zealand ETS), South Korea (Korea ETS), Switzerland (Switzerland ETS) and the United States (Regional Greenhouse Gas Initiative (RGGI) and California Cap-and-Trade (CAT)).[3] Among them, China, the world's largest emitter of GHG, is the first developing economy that has officially run its ETS for carbon reductions. Since 2013, eight pilot carbon trading markets in Beijing, Tianjin, Shanghai, Hubei, Guangdong, Chongqing, Shenzhen, and Fujian have been established. A national carbon trading market, which is the world's biggest, was officially launched in July 2021. The newly established market covers 2,225 domestic power operators using fossil fuels, accounting for more than 40% of the country's total CO_2 emissions and 14% of fossil-fuel-related CO_2 emissions in the world.

Evaluating the impact of existing ETS provides empirical evidence for designing and refining policy instruments for achieving net-zero emissions. Studies on this issue mainly focused on the EU ETS (e.g., De Perthuis and Trotignon 2014; Bel and Joseph 2015; Martin et al. 2016; Chèze et al. 2020; Quemin 2022). Some researchers evaluated RGGI (e.g., Fell and Maniloff 2018; Yan 2021), California CAT (Shen et al. 2014; Caron et al. 2015), New Zealand ETS (Leining et al. 2020), and Korea ETS (Choi et al. 2017). Noteworthily, all those trading systems are run in developed economies, which have relatively well-functioned markets and government regulations. Lessons from them might not be able to directly guide development of ETS and future policy in the context of developing economies. However, empirical evidence on investigating the impact of ETS in developing economies is still limited, partly due to the fact that the majority of developing economies have not established their ETS for controlling carbon emissions.

---

[3] https://ec.europa.eu/clima/eu-action/eu-emissions-trading-system-eu-ets/international-carbon-market_en.

This chapter aims to empirically investigate the impact of China's pilot carbon emission trading, the world's first ETS established in a developing context for controlling carbon emissions. Considering that cities are responsible for the majority of carbon emissions in China (Dhakal 2009; Chen et al. 2021; Wu et al. 2021a, b), the analysis is conducted from the perspective of urban carbon emissions. The propensity score matching-difference-in-differences (PSM-DID) approach is applied to analyse a city-level panel dataset covering the 2010–2016 period. Potential channels underlying the identified relationship have also been explored. We hope that our analysis could provide a reference for implementing ETS in other developing economies.

7.2 Methodology

7.2.1 The Propensity Score Matching-Difference-In-Differences (PSM-DID) Approach

Several recent studies analysed the effects of China's pilot carbon trading markets applying the difference-in-differences (DID) approach based on provincial level panel data (Dong et al. 2019; Zhang et al. 2019; Yang et al. 2021). Noticeably, the effective use of DID approach asks samples to meet the common trend and randomness assumptions (Lechner 2010; Tan et al. 2018; Chen et al. 2022), implying that applying this method directly might cause endogenous problems and biased results. In terms of China's pilot carbon emission trading, the selection of pilot regions was decided by the central government with the consideration of regional balances, local economic development level, local carbon emissions profile and other socioeconomic factors (Table 7.1) (Zhang 2015).[4] Zhang et al. (2014) overviewed the initial status of China's pilot carbon emission trading and found that most of the pilot regions are relatively affluent regions with carbon emissions intensities lower than the country's average. Therefore, the pilot regions were not selected randomly.

To overcome the above bias issue, the PSM-DID approach is used in this study. The PSM-DID approach is a combination of the DID and PSM methods. The PSM, a nonparametric method proposed by Rosenbaum and Rubin (1983), simulates a controlled experiment process for non-randomly selected groups (Bilicka 2019). Individuals who differ in certain indicators could be close matches according to their propensity to be treated. The PSM-DID approach is therefore able to correct potential selection bias caused by observable characteristics. In our case, the treatment is the pilot ETS. Sample cities are divided into two groups: the pilot cities as the treatment group and the non-pilot cities as the control group. The propensity score, which measures the conditional probability of receiving a treatment, is estimated as follows:

---

[4] https://zfxxgk.ndrc.gov.cn/web/iteminfo.jsp?id=1349.

Table 7.1 Overview of China's pilot regions

Region	Population (10^4)	GDP (10^9 yuan)	Energy consumption (10^9 tons standard coal)	Energy intensity (Tons standard coal/10^4 yuan)
Beijing	1,961	1,411	70	0.5
Shanghai	2,304	1,717	112	0.7
Guangdong	10,430	4,601	272	0.7
Shenzhen	1,037	959	49	0.5
Tianjin	1,294	922	68	0.8
Hubei	5,724	1,598	151	1.2
Chongqing	2,885	793	71	1.1

Source Zhang et al. (2017). Year 2010 data is used in this table

$$p(X_i) = p(Z_i = 1|X_i) = E(Z_i|X_i) = f(X_i). \qquad (7.1)$$

where $p(X_i)$ denotes the propensity score, Z_i is a dummy variable which equals 1 if city i is in the treatment group and 0 otherwise, X_i denotes the matrix of observable characteristics of city i, E denotes a mathematical expectation. $p(X_i)$ delivers the likelihood of city i being influenced by the pilot ETS with X_i.

To well match the treatment and the control groups, Kernel matching method is used as Ren et al. (2018) and Ouyang et al. (2020) suggested. More details about the implementation steps of PSM can be found in Caliendo and Kopeinig (2008). After the process, confoundedness and significant differences between the treatment and control groups have been eliminated.

Then, a DID model is established to assess the average effect of China's pilot carbon emission trading on urban carbon emissions.

$$UE_{it} = \alpha_0 + \beta_0 \cdot Treat \cdot time + \beta_1 \cdot X_{it} + f_i + f_t + \varepsilon_{it} \qquad (7.2)$$

where UE_{it} denotes the carbon emissions profile of city i in year t. α_0 is the constant term. $Treat$ is a dummy variable which equals 1 if city i is in the treatment group and 0 otherwise, $time$ is a time dummy variable which equals 1 if $t \geq 2014$ and 0 otherwise. β_0 is the DID estimator measuring the effect of pilot carbon emission trading on urban carbon emissions. X_{it} denotes the matrix of control variables which might impact the carbon intensity of city i. f_i, f_t and ε_{it} denote the region-fixed effect, the year-fixed effect and random errors, respectively.

This chapter also wants to explore the dynamic marginal effect of pilot carbon emission trading on urban carbon emissions. The above DID model is extended with additional time dummy variables as follows:

$$UE_{it} = \alpha + \beta_2 \cdot Treat \cdot t_{2015} + \beta_3 \cdot Treat \cdot t_{2016} + \beta_4 \cdot X_{it} + f_i + f_t + \varepsilon_{it} \qquad (7.3)$$

where t_{2015} and t_{2016} are time dummy variables. $t_{2015} = 1$ if $t = 2015$ and $t_{2016} = 1$ if $t = 2016$. Otherwise, they equal 0. β_2 and β_3 measure the marginal effects of pilot carbon emission trading in 2015 and 2016, respectively.

7.2.2 Variables and Data

It should be noted that China's pilot carbon emission trading, different from ETS schemes operating in other economies such as the EU ETS, RGGI and New Zealand ETS, was set based on carbon intensity reduction goals, rather than absolute emissions reduction. Therefore, this study uses carbon intensity to reflect the urban emissions level. Specifically, the log value of urban carbon emissions is used to represent UE_{it}. Since China's city-level carbon emissions data are not publicly accessible, this chapter estimates city-level carbon emissions using compiled urban energy consumption data according to Wu and Ma (2019), Fu et al. (2021) and Wu et al. (2021b). Details about the carbon emissions estimation method can be found in Wu and Ma (2019) and Wu et al. (2021b). The relevant data for the emissions estimation are from *China City Statistical Yearbook* (NBSC 2011–2017a), *China Energy Statistical Yearbook* (NBSC 2011–2017c) and *China Electric Power Yearbook* (NBSC 2011–2017b).

Existing literature has shown that energy use, economic development level, industry structure and population size might impact urban carbon intensity (Jorgenson 2014; Cui et al. 2019; Mi et al. 2020; Yu et al. 2020; Tang et al. 2021b). Therefore, this chapter controls these factors in the empirical analysis. Specifically, energy use (*lneu*) is measured by electricity consumption per unit of gross city product (GCP) (kW·h/Yuan, log value). Economic development level (*lnw*) is measured by the mean wage (Yuan, log value). Industry structure (*lnser*) is measured by the share of service industries in GCP (%, log value). Population size (*lnp*) is measured by the year-end population (10^4, log value). Besides, this chapter also considers GCP per capita (*lngcp*, Yuan per capita, log value), the share of manufacturing industries in GCP (*lnmanu*) (%, log value) and the share of official R&D investment in GCP (*lnrd*) (%, log value) since the selection of pilot regions in China was potentially influenced by local economic development level, carbon emissions profile and other socioeconomic factors. The data used for measuring those variables are from *China City Statistical Yearbook* (NBSC 2011–2017a), *China Energy Statistical Yearbook* (NBSC 2011–2017c), *China Electric Power Yearbook* (NBSC 2011–2017b) and *China Statistical Yearbook* (NBSC 2011–2017d).

In October 2011, China's government firstly selected five cities (Beijing, Tianjin, Shanghai, Shenzhen and Chongqing) and two provinces (Hubei and Guangdong) as regions for establishing pilot ETS. Seven pilot ETS schemes were officially launched in those regions in 2013–2014. In late 2016, the eighth pilot ETS scheme was launched in the province of Fujian. This chapter chooses the 2010–2016 period for the analysis. For the purpose of simplicity, 2014 is set as the year when the pilot ETS schemes were launched. Prefecture-level cities in the seven pilot regions are considered as the treatment group, while other prefecture-level cities in China's mainland are considered as the control group. The initial treatment and control groups include 34 and 239 cities, respectively.

7.3   Results and Discussion

7.3.1   Results of the Propensity Score Matching (PSM) Analysis

A PSM analysis is conducted to evaluate the initial treatment and control groups. Kernel matching method is used to well match the treatment and the control groups. Probit regression is used to estimate the propensity scores. After the treatment of PSM, the treatment and the control groups include 30 and 200 cities, respectively. 4 cities in the initial treatment group and 39 cities in the initial control group are excluded.

Table 7.2 compare the identified variables, including energy use (*lneu*), economic development level (*lnw*), industry structure (*lnser*), population size (*lnp*), GCP per capita (*lngcp*), the share of manufacturing industries in GCP (*lnmanu*) and the share of official R&D investment in GCP (*lnrd*), before and after the conducted PSM analysis. After the PSM treatment, the identified variables' standard deviations decline substantially. In addition, the insignificant results of the *t* values imply that no significant differences exist after matching. Moreover, the kernel densities of treatment and the control groups are much closer after matching. These confirm that the method applied is reasonable and effective. After the matching, all variables turn out to be balanced and the residual difference between the treatment and the control groups has been eliminated. The identified variables in the new control group are not systematically different from those in the treatment group.

7.3.2   Results of the DID Analysis

The key precondition of adopting the DID model is that there is a common change trend between the treatment and control groups before the policy; that is, the trends should be parallel (Zhang and Wang 2021). Figure 7.1 plots annual mean urban carbon intensity trends for the two groups. It shows that the mean trends of two groups were similar before the pilot ETS was implemented, conforming to the parallel trend requirement. After the launch of the pilot ETS, there was clear difference in carbon intensity between the two groups. Cities in the treatment group tended to have more rapidly declining carbon intensities than those in the control group.

Table 7.2 Balance test of variables

Variable	Unmatched/Matched	Treatment group mean	Control group mean	S.D	t value	P-value
lneu	U	1.36	1.52	−19.7	−1.10	0.27
	M	1.39	1.39	−0.5	−0.02	0.98
lnw	U	10.60	10.42	52.6	3.30	0.00
	M	10.50	10.48	7.2	0.40	0.69
lnser	U	3.65	3.55	41.1	2.20	0.03
	M	3.61	3.59	8.5	0.37	0.71
lnp	U	6.04	5.86	26.0	1.41	0.16
	M	5.97	6.00	3.9	−0.16	0.88
lngcp	U	10.71	10.30	51.4	3.13	0.00
	M	10.50	10.44	7.5	0.35	0.73
lnmanu	U	3.87	3.92	−26.5	−1.32	0.19
	M	3.89	3.90	−5.9	−0.24	0.82
lnrd	U	−1.54	−1.65	16.5	0.97	0.33
	M	−1.65	−1.61	−6.4	−0.27	0.79

Note S.D. notes standard deviation

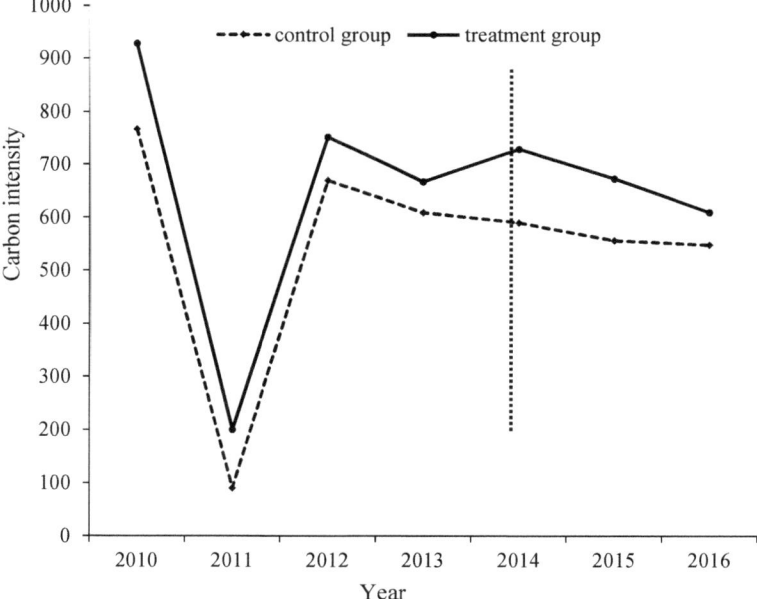

Fig. 7.1 Annual mean urban carbon intensity

Table 7.3 The average effect of pilot ETS on urban carbon emissions

Variable	Average treatment effect	
	(1)	(2)
$Treat \cdot time$	−0.85***(−7.32)	−0.78***(−7.34)
lneu		1.33***(10.60)
lnser		0.67(1.29)
lnw		$-_{0.04}(-0.38)$
lnp		−0.02(−0.045)
cons	6.31***(197.43)	2.53(0.70)
Year-fixed effect	Yes	Yes
Region-fixed effect	Yes	Yes
R^2	0.12	0.30

Note The values in the brackets are the t values of two-tailed tests. ***, **, and * denote significance levels at the 1%, 5%, and 10%, respectively

Then, the established DID model (Eq. 7.2) is used to assess the average effect of pilot ETS on urban carbon emissions (Table 7.3). Colum (1) of Table 7.3 presents the results of basic regression. Column (2) presents the results with control variables. The coefficient of $Treat \cdot time$ remains significantly negative in both regressions, indicating that the pilot ETS has generally reduced urban carbon emissions of cities covered by the scheme. With pilot ETS, the average urban carbon intensity of cities covered by the pilot ETS is about 80% lower than that of the control group. Besides, only the coefficient of energy use is significant, arguing that energy use was positively associated with urban carbon intensity.

Table 7.4 report the dynamic marginal effect results. Whether control variables are included or not, the coefficients of $Treat \cdot t_{2015}$ and $Treat \cdot t_{2016}$ remain significantly negative, indicating that the pilot ETS has steadily reduced the carbon intensity of cities covered by the scheme after its implementation. Additionally, the difference between the coefficients of $Treat \cdot t_{2015}$ and $Treat \cdot t_{2016}$ also indicate that the reducing effect in 2016 is larger than that in 2015. Similarly, only the coefficient of energy use is significant which argues that energy use was positively associated with urban carbon intensity.

7.3.3 Robustness Tests

To evaluate the credibility of the above empirical analysis, two robustness testes have been conducted. First, additional control variables are introduced. There are noticeable disparities across China's cities in diverse aspects which potentially influenced the selection of pilot regions. To better reflect those disparities and eliminate selection bias, this chapter introduces additional three control variables including GCP

Table 7.4 The dynamic marginal effect of pilot ETS on urban carbon emissions

Variable	Dynamic marginal effect	
	(1)	(2)
$Treat \cdot t_{2015}$	$-0.62^{***}(-4.64)$	$-0.59^{***}(-4.90)$
$Treat \cdot t_{2016}$	$-1.09^{***}(-8.20)$	$-0.97^{***}(-8.03)$
lneu		$1.31^{***}(10.48)$
lnser		$0.57(1.10)$
lnw		$-0.04(-0.34)$
lnp		$-0.02(-0.04)$
cons	6.31^{***}	2.88
	(199.92)	(0.80)
Year-fixed effect	Yes	Yes
Region-fixed effect	Yes	Yes
R^2	0.20	0.32

Note The values in the brackets are the *t* values of two-tailed tests. ***, **, and * denote significance levels at the 1%, 5%, and 10%, respectively

per capita (*lngcp*), the share of manufacturing industries in GCP (*lnmanu*) and the share of official R&D investment in GCP (*lnrd*). The result shows that introducing more control variables has little impact on the coefficient of $Treat \cdot time$ (Table 7.5).

Table 7.5 Robustness test: additional control variables

Variable	(1)
$Treat \cdot time$	$-0.82^{***}(-7.44)$
lneu	$1.34^{***}(10.55)$
lnser	$1.20(1.57)$
lnw	$-0.10(-0.80)$
lnp	$-0.88(-1.35)$
lngcp	$-0.88^{*}(-1.67)$
lnrd	$0.06(1.07)$
lnrd	$0.06(1.07)$
lnmanu	$1.12(1.3)$
cons	$11.21(1.19)$
Region-fixed effect	Yes
Year-fixed effect	Yes
R^2	0.30

Note The values in the brackets are the *t* values of two-tailed tests. ***, **, and * denote significance levels at the 1%, 5%, and 10%, respectively

Table 7.6 Robustness test: counterfactual test

Variable	(1)	(2)	(3)
$Treat \cdot time$	0.11(1.04)	0.07(0.67)	0.07(0.67)
lneu	1.38***(10.42)	1.38***(10.38)	1.38***(10.38)
lnser	1.18**(2.16)	1.20**(2.19)	1.21**(2.21)
lnw	−0.02(−0.14)	−0.02(−0.20)	−0.02(−0.19)
lnp	−0.12(−0.24)	−0.12(−0.25)	−0.17(−0.35)
cons	0.92(0.24)	0.96(0.25)	1.23(0.32)
Region-fixed effect	Yes	Yes	Yes
Year-fixed effect	Yes	Yes	Yes
R^2	0.22	0.22	0.22

Note The values in the brackets are the t values of two-tailed tests. ***, **, and * denote significance levels at the 1%, 5%, and 10%, respectively

In addition, a counterfactual test is conducted. This chapter randomly selects 30 cities as dummy pilot samples and repeats the DID analysis. If the result does not support that the pilot ETS has reduced the carbon intensity of dummy pilot cities, the previous results should be robust. Otherwise, the previous results would be questioned. Three groups of dummy pilot samples are randomly selected and the results are shown in Table 7.6. The coefficient of $Treat \cdot time$ becomes insignificant in all regressions, saying that the pilot ETS did not significantly influence dummy pilot samples. Overall, the conducted robustness tests support the credibility of the above empirical results.

7.3.4 Mechanisms

This chapter also examines potential channels. A mediating analysis is conducted to explain the observed effects on urban carbon intensity. Details about the mediating method used can be found in Saeidi et al. (2015) and Tang et al. (2020b). Industry structure (*lnser*), energy use (*lneu*) and R&D investment (*lnrd*) are identified as potential mediators. The results are presented in Table 7.7. The analysis provides evidence that adjusting industrial structure and decreasing energy intensity can explain the observed effects on urban carbon intensity. However, increasing R&D investment would supress, rather than mediate, the effect of the pilot ETS on urban carbon intensity (Columns 6 and 7).

Though it is well documented that promoting R&D plays a key role in reducing carbon emissions, some researchers have noticed that increasing R&D investment may result in rising carbon intensity under certain contexts. Churchill et al. (2019) analysed longitudinal data from G7 countries and highlighted that the production scale effect associated with growth and trade openness is the underlying cause; that

Table 7.7 Mediating analysis

Variable	(1)	(2)	(3)	(4)	(5)	(6)	(7)
	UE	lnser	UE	lneu	UE	lnrd	UE
$Treat \cdot time$	−0.11***	0.03***	−0.09***				
	(−3.16)	(2.58)	(−2.69)				
lnser			−0.52***				
			(−4.15)				
$Treat \cdot time$	−0.11***			−0.05***	−0.07*		
	(-3.16)			(−4.64)	(−1.90)		
lneu					0.92***		
					(71.46)		
$Treat \cdot time$	−0.11***					0.34***	−0.16***
	(−3.16)					(4.09)	(−3.48)
lnrd							0.04***
							(1.98)
Control variable	Yes	Yes	Yes	Yes	Yes	Yes	Yes
Region-fixed effect	Yes	Yes	Yes	Yes	Yes	Yes	Yes
Year-fixed effect	Yes	Yes	Yes	Yes	Yes	Yes	Yes
R^2	0.04	0.26	0.07	0.01	0.92	0.11	0.04

Note The values in the brackets are the *t* values of two-tailed tests. ***, **, and * denote significance levels at the 1%, 5%, and 10%, respectively

is, expanding output still needs using natural resources even with increasing R&D investment, thus emitting more emissions inevitably. Potential energy rebound effect may also contribute to this positive relationship (Jia et al. 2019; Bai et al. 2020). Similar positive relationship has also been observed in several European countries, Canada, Turkey, the United States and some regions in China (Song et al. 2019; Petrović and Lobanov 2020).

7.4 Conclusions

This chapter empirically investigates the impact of China's pilot carbon emission trading, the world's first ETS established in a developing context for controlling carbon emissions, on urban carbon emissions. The PSM-DID approach is applied to analyse a city-level panel dataset covering the 2010–2016 period. In addition, this chapter examines potential channels underlying the relationship between the pilot ETS and urban carbon emissions.

The results show that the pilot ETS has reduced urban carbon emissions of cities covered by the scheme. With pilot ETS, the average urban carbon intensity of cities covered by the pilot ETS is about 80% lower than that of the control group. When examining potential channels, the analysis provides evidence that adjusting industrial structure and decreasing energy intensity can explain the observed reduction effect. However, increasing R&D investment would supress, rather than mediate, the effect of the pilot ETS. Overall, the pilot ETS has played a positive role in promoting the country's carbon stabilization.

Future research could extend this analysis by exploring firm-level data to reveal the micro mechanism of China's pilot ETS on carbon emissions. The policy spillover effect of carbon ETS also needs further investigation.

References

Aldy, J. E., & Stavins, R. N. (2012). The promise and problems of pricing carbon: Theory and experience. *Journal of Environment & Development*, 21(2), 152-180.

Andresen, S., Bang, G., Skjærseth, J. B., & Underdal, A. (2021). Achieving the ambitious targets of the Paris Agreement: The role of key actors. *International Environmental Agreements: Politics, Law and Economics*, 21(1), 1-7.

Bai, C., Feng, C., Yan, H., Yi, X., Chen, Z., & Wei, W. (2020). Will income inequality influence the abatement effect of renewable energy technological innovation on carbon dioxide emissions?. *Journal of Environmental Management*, 264, 110482.

Bakam, I., Balana, B. B., & Matthews, R. (2012). Cost-effectiveness analysis of policy instruments for greenhouse gas emission mitigation in the agricultural sector. *Journal of Environmental Management*, 112, 33-44.

Bel, G., & Joseph, S. (2015). Emission abatement: Untangling the impacts of the EU ETS and the economic crisis. *Energy Economics*, 49, 531-539.

Bilicka, K. A. (2019). Comparing UK tax returns of foreign multinationals to matched domestic firms. *American Economic Review*, 109(8), 2921-53.

Borghesi, S., Cainelli, G., & Mazzanti, M. (2015). Linking emission trading to environmental innovation: Evidence from the Italian manufacturing industry. *Research Policy*, 44(3), 669-683.

Bullock, D. (2012). Emissions trading in New Zealand: Development, challenges and design. *Environmental Politics*, 21(4), 657-675.

Caliendo, M., & Kopeinig, S. (2008). Some practical guidance for the implementation of propensity score matching. *Journal of Economic Surveys*, 22(1), 31-72.

Caron, J., Rausch, S., & Winchester, N. (2015). Leakage from sub-national climate policy: The case of California's cap–and–trade program. *The Energy Journal*, 36(2), 167-190.

Chen, X., Chen, G., Lin, M., Tang, K., & Ye, B. (2022). How does anti-corruption affect enterprise green innovation in China's energy-intensive industries?. *Environmental Geochemistry and Health*, 44, 2919-2942.

Chen, J., Gao, M., Cheng, S., Liu, X., Hou, W., Song, M., Li, D., & Fan, W. (2021). China's city-level carbon emissions during 1992–2017 based on the inter-calibration of nighttime light data. *Scientific Reports*, 11(1), 1-13.

Chèze, B., Chevallier, J., Berghmans, N., & Alberola, E. (2020). On the CO_2 emissions determinants during the EU ETS Phases I and II: A plant-level analysis merging the EUTL and Platts power data. *The Energy Journal*, 41(4), 153-184.

Choi, Y., Liu, Y., & Lee, H. (2017). The economy impacts of Korean ETS with an emphasis on sectoral coverage based on a CGE approach. *Energy Policy*, 109, 835-844.

Churchill, S. A., Inekwe, J., Smyth, R., & Zhang, X. (2019). R&D intensity and carbon emissions in the G7: 1870–2014. *Energy Economics, 80*, 30-37.

Cui, P., Xia, S., & Hao, L. (2019). Do different sizes of urban population matter differently to CO_2 emission in different regions? Evidence from electricity consumption behavior of urban residents in China. *Journal of Cleaner Production, 240*, 118207.

Dhakal, S. (2009). Urban energy use and carbon emissions from cities in China and policy implications. *Energy Policy, 37*(11), 4208-4219.

De Perthuis, C., & Trotignon, R. (2014). Governance of CO_2 markets: Lessons from the EU ETS. *Energy Policy, 75*, 100-106.

Dong, F., Dai, Y., Zhang, S., Zhang, X., & Long, R. (2019). Can a carbon emission trading scheme generate the Porter effect? Evidence from pilot areas in China. *Science of The Total Environment, 653*, 565-577.

Fankhauser, S., Smith, S. M., Allen, M., et al. (2022). The meaning of net zero and how to get it right. *Nature Climate Change, 12*(1), 15-21.

Fell, H., & Maniloff, P. (2018). Leakage in regional environmental policy: The case of the regional greenhouse gas initiative. *Journal of Environmental Economics and Management, 87*, 1-23.

Fu, Y., He, C., & Luo, L. (2021). Does the low-carbon city policy make a difference? Empirical evidence of the pilot scheme in China with DEA and PSM-DID. *Ecological Indicators, 122*, 107238.

Jia, J., Jian, H., Xie, D., Gu, Z., & Chen, C. (2019). Multi-scale decomposition of energy-related industrial carbon emission by an extended logarithmic mean Divisia index: A case study of Jiangxi, China. *Energy Efficiency, 12*(8), 2161-2186

Jorgenson, A. K. (2014). Economic development and the carbon intensity of human well-being. *Nature Climate Change, 4*(3), 186-189.

Lechner, M. (2010). The estimation of causal effects by difference-in-difference methods. Discussion paper, Economics Department, University of St. Gallen.

Leining, C., Kerr, S., & Bruce-Brand, B. (2020). The New Zealand Emissions Trading Scheme: Critical review and future outlook for three design innovations. *Climate Policy, 20*(2), 246-264.

Martin, R., Muûls, M., & Wagner, U. J. (2016). The impact of the European Union Emissions Trading Scheme on regulated firms: what is the evidence after ten years?. *Review of Environmental Economics and Policy, 10*(1), 129-48.

Meinshausen, M., Meinshausen, N., Hare, W., Raper, S. C., Frieler, K., Knutti, R., Frame, D.J., & Allen, M. R. (2009). Greenhouse-gas emission targets for limiting global warming to 2 °C. *Nature, 458*(7242), 1158-1162.

Meleo, L., Nava, C. R., & Pozzi, C. (2016). Aviation and the costs of the European Emission Trading Scheme: The case of Italy. *Energy Policy, 88*, 138-147.

Mi, Z., Zheng, J., Meng, J., Ou, J., Hubacek, K., Liu, Z., Coffman, D.M., Stern, N., Liang, S., & Wei, Y. M. (2020). Economic development and converging household carbon footprints in China. *Nature Sustainability, 3*(7), 529-537.

Michaelowa, A., Allen, M., & Sha, F. (2018). Policy instruments for limiting global temperature rise to 1.5° C–Can humanity rise to the challenge?. *Climate Policy, 18*(3), 275–286.

Miola, A., Marra, M., & Ciuffo, B. (2011). Designing a climate change policy for the international maritime transport sector: Market-based measures and technological options for global and regional policy actions. *Energy Policy, 39*(9), 5490-5498.

Narassimhan, E., Gallagher, K. S., Koester, S., & Alejo, J. R. (2018). Carbon pricing in practice: A review of existing emissions trading systems. *Climate Policy, 18*(8), 967-991.

NBSC (National Bureau of Statistics of China) (2011–2017a). *China City Statistical Yearbook.* Beijing: China Statistics Press.

NBSC (2011–2017b). *China Electric Power Statistical Yearbook.* Beijing: China Electric Power Press.

NBSC (2011–2017c). *China Energy Statistical Yearbook.* Beijing: China Statistics Press.

NBSC (2011–2017d). *China Statistical Yearbook.* Beijing: China Statistic Press.

Nieto, J., Carpintero, Ó., & Miguel, L. J. (2018). Less than 2°C? An economic-environmental evaluation of the Paris Agreement. *Ecological Economics*, 146, 69-84.

Ouyang, X., Fang, X., Cao, Y., & Sun, C. (2020). Factors behind CO_2 emission reduction in Chinese heavy industries: Do environmental regulations matter?. *Energy Policy*, 145, 111765.

Peake, S., & Ekins, P. (2017). Exploring the financial and investment implications of the Paris Agreement. *Climate Policy*, 17(7), 832-852.

Petrović, P., & Lobanov, M. M. (2020). The impact of R&D expenditures on CO_2 emissions: Evidence from sixteen OECD countries. *Journal of Cleaner Production*, 248, 119187.

Quemin, S. (2022). Raising climate ambition in emissions trading systems: The case of the EU ETS and the 2021 review. *Resource and Energy Economics*, 68, 101300.

Ranson, M., & Stavins, R. N. (2016). Linkage of greenhouse gas emissions trading systems: Learning from experience. *Climate Policy*, 16(3), 284-300.

Ren, H., Folmer, H., & Van der Vlist, A. J. (2018). The impact of home ownership on life satisfaction in urban China: A propensity score matching analysis. *Journal of Happiness Studies*, 19(2), 397-422.

Roberts, C., Geels, F. W., Lockwood, M., Newell, P., Schmitz, H., Turnheim, B., & Jordan, A. (2018). The politics of accelerating low-carbon transitions: Towards a new research agenda. *Energy Research & Social Science*, 44, 304-311.

Rogelj, J., Den Elzen, M., Höhne, N., Fransen, T., Fekete, H., Winkler, H., Schaeffer, R., Sha, F., Riahi, K., & Meinshausen, M. (2016). Paris Agreement climate proposals need a boost to keep warming well below 2°C. *Nature*, 534(7609), 631-639.

Rogge, K. S., Schneider, M., & Hoffmann, V. H. (2011). The innovation impact of the EU Emission Trading System: Findings of company case studies in the German power sector. *Ecological Economics*, 70(3), 513-523.

Rosenbaum, P. R., & Rubin, D. B. (1983). The central role of the propensity score in observational studies for causal effects. *Biometrika*, 70(1), 41–55.

Sachs, J. D., Schmidt-Traub, G., & Williams, J. (2016). Pathways to zero emissions. *Nature Geoscience*, 9(11), 799-801.

Saeidi, S. P., Sofian, S., Saeidi, P., Saeidi, S. P., & Saaeidi, S. A. (2015). How does corporate social responsibility contribute to firm financial performance? The mediating role of competitive advantage, reputation, and customer satisfaction. *Journal of Business Research*, 68(2), 341-350.

Schneider, L., Lazarus, M., Lee, C., & Van Asselt, H. (2017). Restricted linking of emissions trading systems: Options, benefits, and challenges. *International Environmental Agreements: Politics, Law and Economics*, 17(6), 883-898.

Shen, B., Dai, F., Price, L., & Lu, H. (2014). California's cap-and-trade programme and insights for China's pilot schemes. *Energy & Environment*, 25(3-4), 551-575.

Song, X., Zhou, Y., & Jia, W. (2019). How do economic openness and R&D investment affect green economic growth? Evidence from China. *Resources, Conservation and Recycling*, 146, 405-415.

Tang, K., & Ma, C. (2022). The cost-effectiveness of agricultural greenhouse gas reduction under diverse carbon policies in China. *China Agricultural Economic Review*. https://doi.org/10.1108/CAER-01-2022-0008

Tan, R., Tang, D., & Lin, B. (2018). Policy impact of new energy vehicles promotion on air quality in Chinese cities. *Energy Policy*, 118, 33-40.

Tang, K., Hailu, A., & Yang, Y. (2020a). Agricultural chemical oxygen demand mitigation under various policies in China: A scenario analysis. *Journal of Cleaner Production*, 250, 119513.

Tang, K., Qiu, Y., & Zhou, D. (2020b). Does command-and-control regulation promote green innovation performance? Evidence from China's industrial enterprises. *Science of the Total Environment*, 712, 136362.

Tang, K., Wang, M., & Zhou, D. (2021a). Abatement potential and cost of agricultural greenhouse gases in Australian dryland farming system. *Environmental Science and Pollution Research*, 28(17), 21862-21873.

Tang, K., Xiong, C., Wang, Y., & Zhou, D. (2021b). Carbon emissions performance trend across Chinese cities: Evidence from efficiency and convergence evaluation. *Environmental Science and Pollution Research*, 28(2), 1533-1544.

Tang, K., Zhou, Y., Liang, X., & Zhou, D. (2021c). The effectiveness and heterogeneity of carbon emissions trading scheme in China. *Environmental Science and Pollution Research*, 28(14), 17306-17318.

van Soest, H. L., den Elzen, M. G., & van Vuuren, D. P. (2021). Net-zero emission targets for major emitting countries consistent with the Paris Agreement. *Nature Communications*, 12(1), 1-9.

Verbruggen, A., Laes, E., & Woerdman, E. (2019). Anatomy of emissions trading systems: What is the EU ETS?. *Environmental Science & Policy*, 98, 11-19.

Wu, J., & Ma, C. (2019). The convergence of China's marginal abatement cost of CO_2: an emission-weighted continuous state space approach. *Environmental and Resource Economics*, 72(4), 1099-1119.

Wu, J., Feng, Z., & Tang, K. (2021a). The dynamics and drivers of environmental performance in Chinese cities: A decomposition analysis. *Environmental Science and Pollution Research*, 28(24), 30626-30641.

Wu, J., Xu, H., & Tang, K. (2021b). Industrial agglomeration, CO_2 emissions and regional development programs: A decomposition analysis based on 286 Chinese cities. *Energy*, 225, 120239.

Wu, Q., & Wang, Y. (2022). How does carbon emission price stimulate enterprises' total factor productivity? Insights from China's emission trading scheme pilots. *Energy Economics*, 109, 105990.

Yan, J. (2021). The impact of climate policy on fossil fuel consumption: Evidence from the Regional Greenhouse Gas Initiative (RGGI). *Energy Economics*, 100, 105333.

Yang, L., Li, Y., & Liu, H. (2021). Did carbon trade improve green production performance? Evidence from China. *Energy Economics*, 96, 105185

Yao, S., Yu, X., Yan, S., & Wen, S. (2021). Heterogeneous emission trading schemes and green innovation. *Energy Policy*, 155, 112367.

Yu, X., Wu, Z., Zheng, H., Li, M., & Tan, T. (2020). How urban agglomeration improve the emission efficiency? A spatial econometric analysis of the Yangtze River Delta urban agglomeration in China. *Journal of Environmental Management*, 260, 110061.

Zhang, D., Karplus, V. J., Cassisa, C., & Zhang, X. (2014). Emissions trading in China: Progress and prospects. *Energy Policy*, 75, 9-16.

Zhang, M., Liu, Y., & Su, Y. (2017). Comparison of carbon emission trading schemes in the European Union and China. *Climate*, 5(3), 70.

Zhang, W., Zhang, N., & Yu, Y. (2019). Carbon mitigation effects and potential cost savings from carbon emissions trading in China's regional industry. *Technological Forecasting and Social Change*, 141, 1-11.

Zhang, Y. J., & Wang, W. (2021). How does China's carbon emissions trading (CET) policy affect the investment of CET-covered enterprises?. *Energy Economics*, 98, 105224.

Zhang, Z. (2015). Carbon emissions trading in China: The evolution from pilots to a nationwide scheme. *Climate Policy*, 15(sup1), S104-S126.

Zhou, D., Liang, X., Zhou, Y., & Tang, K. (2020). Does emission trading boost carbon productivity? Evidence from China's pilot emission trading scheme. *International Journal of Environmental Research and Public Health*, 17(15), 5522.

Chapter 8
Investigating the Impact of Carbon Emission Trading on Provincial Industrial Carbon Emissions in China

Kai Tang and Ye Zhou

8.1 Introduction

The Paris Agreement adopted in 2015 has prompted global policymakers to consider more effective policy tools for moving towards net-zero emissions. There is widespread agreement among most researchers that carbon pricing is likely to be an economically effective instrument which should play a key role in efficient greenhouse gases (GHG) reduction and mitigation (Brauneis et al. 2013; Tang et al. 2016a, b; Boyce 2018; Klenert et al. 2018; Skovgaard et al. 2019; van den Bergh and Botzen 2020; Mildenberger et al. 2022; Stavins 2022). In practice, carbon pricing could occur via carbon taxation or emissions trading system (ETS). To date, there are 64 carbon pricing schemes in operation and three planned for establishment across the world (Fig. 8.1), covering approximately 22% of global GHG emissions (World Bank 2021).

Instead of commanding who must mitigate GHG emissions where and how, carbon pricing such as an ETS creates a financial incentive to mitigate emissions through delivering a clear economic signal. Then, emitters can choose their own solution, either taking actions to reduce their emissions or continue emitting and paying for the externality inherent to their emissions, depending on the relative costs of those choices (Pigou 1932; Coase 1960; Stavins 2022). With effective carbon pricing, the society's GHG emissions goal is achieved in a flexible and cost-effective way (Stavins 2008; Boyce 2018; van den Bergh and Botzen 2020; Tang and Ma 2022). Such pricing shifts the burden for the external costs of GHG emissions away from society to those who are responsible for it and who can avoid it; that is, carbon pricing

K. Tang (✉)
School of Economics and Trade, Guangdong University of Foreign Studies, Guangzhou 510006, China
e-mail: francistang1988@hotmail.com

Y. Zhou
School of Business, Guangdong University of Foreign Studies, Guangzhou 510006, China

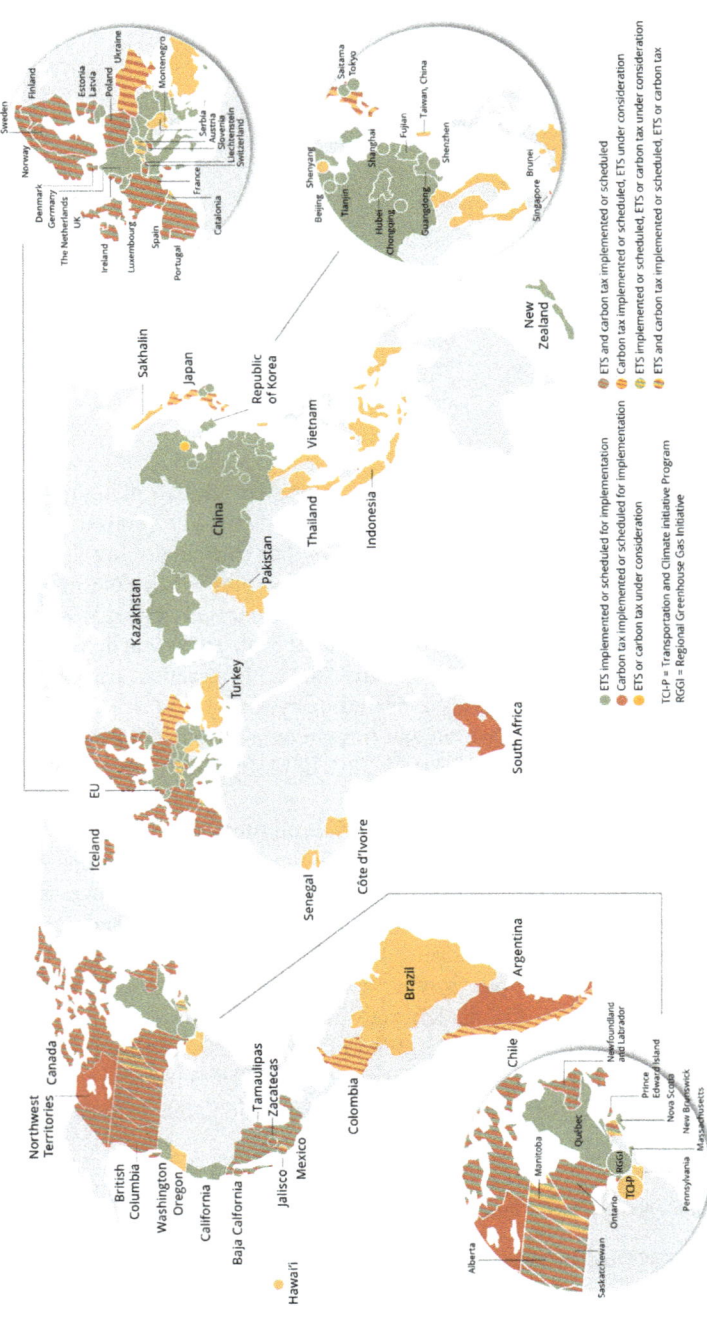

Fig. 8.1 Carbon pricing schemes across the world, 2021. *Source* World Bank Group (2021)

can internalize the external costs of GHG emissions (Tang et al. 2018, 2021b; Tang and Hailu 2020). By incorporating climate change costs into economic decision-making, carbon pricing is likely to encourage modifying production and consumption patterns, thereby underpinning decarbonized development.

An ETS is an instrument that allocates property rights for emitters to emit GHG, as emissions permits, and allows them to trade the allocated rights. The quantity of GHG emissions is regulated economically through an ETS. Generally, there are two types of ETS, including cap-and-trade and baseline-and-credit schemes. A cap-and-trade ETS applies an upper limit, or cap, on GHG emissions and emissions permits are allocated based on the identified cap. In a baseline-and-credit ETS, baseline emissions levels are identified for regulated emitters and credits are issued to those who emit less emissions than their baseline levels. The credits can be sold to other emitters who need them to comply with regulations they face. No fixed limit on GHG emissions is set under a baseline-and-credit ETS.[1]

Since the middle of the 2000s, several ETS schemes have been launched. The EU ETS, set up in 2005, is the first international ETS scheme globally. Besides the EU ETS, national or sub-national systems are already operating or under development in Canada (Québec Cap-and-Trade), China, Japan (Japan Voluntary Emission Trading Scheme (JVETS)), New Zealand (New Zealand ETS), South Korea (Korea ETS), Switzerland (Switzerland ETS) and the United States (Regional Greenhouse Gas Initiative (RGGI) and California Cap-and-Trade (CAT)).[2] In 2011, the Chinese government announced its plan to establish seven pilot regional ETS schemes to meet the national climate goals. Since 2013, those announced pilot schemes have been launched, covering selected industries in four provincial-level municipalities (Beijing, Shanghai, Tianjin, and Chongqing), two provinces (Hubei and Guangdong), and one sub-provincial municipality (Shenzhen). In total, those seven pilot ETS schemes cover approximately 11.4% of the country's emissions in 2014 (Zhang et al. 2017; Tang et al. 2021a). Most of the established ETS schemes are cap-and-trade, but the schemes in China, including the pilot ETS and the recently established national scheme, are baseline-and-credit.

Studies of ETS have largely addressed the related economic and social issues on the cases of cap-and-trade ETS schemes. Most of them have focused on the EU ETS, the flagship of EU climate policy regime (e.g., Convery 2009; De Perthuis and Trotignon 2014; Brink et al. 2016; Hintermann et al. 2016; Löschel et al. 2019; Abate et al. 2021; Oesingmann 2022), and some have covered other cases at the regional or national level including RGGI (e.g., Shawhan et al. 2014; Chan and Morrow 2019), California CAT (Shen et al. 2014; Caron et al. 2015), New Zealand ETS (Richter and Mundaca 2013; Diaz-Rainey and Tulloch 2018), and Korea ETS (Choi et al. 2017; Kim and Yu 2018).

A relatively small but growing body of literature has dealt with China's pilot ETS. This literature largely adopts computable general equilibrium (CGE) modelling or

---

[1] https://www.oecd.org/env/tools-evaluation/emissiontradingsystems.htm.

[2] https://ec.europa.eu/clima/eu-action/eu-emissions-trading-system-eu-ets/international-carbon-market_en.

simulation-based modelling (e.g., Li and Jia 2016; Lin and Jia 2018; Tan et al. 2018; Wu and Gong 2021). These models used often have various assumptions made to allow for analysis. Econometric investigations of China's pilot ETS, which circumvent these assumptions and use real data, are less common, though a few exists. Several econometric studies that assessed this baseline-and-credit ETS tend to rely on provincial-level data (Dong et al. 2019; Chen et al. 2021; Yang et al. 2021), which may lead to over-coverage issue since the pilot ETS only covers limited industries (Table 8.1).

In this chapter we empirically investigate the impact of China's pilot ETS on provincial industrial carbon emissions. The analysis differs from several other studies of pilot ETS using provincial level data in that we rely on provincial-industrial level data in order to accurately reflect the actual coverage of pilot ETS. This also allow us to investigate pilot ETS on carbon emissions across regions and industries. Both the difference-in-differences (DID) and difference-in-difference-in-differences (DDD) approaches are applied to analyse a compiled panel dataset covering the 2008–2017 period. We also explore potential channels underlying the identified relationship. We hope that our analysis could provide empirical evidence for implementing and refining relevant ETS schemes in the future.

8.2 Methodology

8.2.1 Difference-In-Differences (DID) Approach

In literature, the DID approach has been widely applied in evaluating environment policy tools in diverse contexts (e.g., Fell and Maniloff 2018; Koch and Mama 2019; Löschel et al. 2019; Wang and Watanabe 2019; Tang et al. 2020; Tian et al. 2020; Fageda and Teixidó 2022). To determine if China's pilot ETS impacted provincial industrial carbon emissions, this chapter investigates a DID model. The set up DID model is one in which we analyse a treatment effect associated with the pilot ETS, and its specific form is given as:

$$E_{ijt} = \alpha_0 + \beta_0 \cdot Treat \cdot time + \beta_1 \cdot X_{it} + f_i + f_j + f_t + \varepsilon_{ijt} \qquad (8.1)$$

E_{ijt} is the carbon emissions of industry j in province i in year t. α_0 is the constant term. $Treat$ is a dummy variable which equals 1 if industry j is in the pilot region and 0 otherwise. $time$ is a time dummy variable which equals 1 if $t \geq 2013$ (when the pilot ETS is in effect) and 0 otherwise. β_0 is the DID estimator measuring the effect of pilot ETS on industrial carbon emissions. X_{it} denotes a set of controls which might impact the carbon emissions of industry i discussed in more detail below. f_i, f_j, f_t and ε_{ijt} denote the region-fixed effect, the industry-fixed effect, the year-fixed effect and random errors, respectively.

Table 8.1 Covered industries in China's pilot ETS schemes

Pilot region	Threshold	Covered industries	Emissions covered (%)	Permits allocation
Beijing	The average emissions in 2009–2011 > 10^4 tons	Industrial and service sectors	40.0	Free
Shanghai	In 2010–2011, emissions > 2×10^4 tons (industrial) 10^4 tons (non-industrial)	Industrial and non-industrial sectors	57.0	Free
Guangdong	2011–2014 emissions > 2×10^4 tons or energy consumption > 10^4 tons of standard coal	Electricity, cement, steel, petrochemical, hotels, restaurants, business	58.0	Mostly free, a small share via auction
Shenzhen	Annual emissions of 3×10^3 tons for enterprises; 10^4 m^2 for public buildings	Industry and construction	40.0	Free
Tianjin	Enterprises and civil buildings that have emitted > 2×10^4 tons since 2009	Electricity, heating, steel, chemical, petrochemical, oil and gas exploration and civil construction	60.0	Free
Hubei	Energy consumption 6×10^4 tons of standard coal	Electricity, heating, metallurgy, steel, chemicals, cement, petrochemicals, automotive glass, chemical fiber, paper, medicine, food and beverage, papermaking	35.9	Free
Chongqing	Emissions > 2×10^4 tons or energy consumption > 10^4 tons of standard coal	Electricity, electrolytic aluminum, ferroalloy, calcium carbide, caustic soda, cement, steel	39.5	Free

Note Information derived from the official documents of the seven pilot ETS

8.2.2 Difference-In-Difference-In-Differences (DDD) Approach

Although the estimated DID model can deliver the impact of the pilot ETS to a certain extent, the estimated impact may still be challenged by the biased issue since the impact levels on the emissions of different industries may vary significantly. Industries directly covered by the pilot ETS are likely to be more influenced by the scheme. In light of this concern, this chapter investigates a DDD model. The DDD approach is built on the premise that one more source of variation helps to better identify the pure impact of a treatment (Bradley et al. 2016; Cai et al. 2016; Tang et al. 2020; Li et al. 2021). In our specifications, non-covered industries in the pilot and non-pilot regions have been identified as the second treatment and control groups, respectively. Since the non-covered industries are not regulated by the pilot ETS, such a practice is expected to exclude the interference of other factors and divest the net impact of the regulation, thus ruling out threats to validity that the initial DID analysis cannot.

Specifically, the DDD model is given as:

$$E_{ijt} = \alpha_1 + \beta_2 \cdot Treat \cdot time \cdot group + \beta_3 \cdot X_{it} + f_{it} + f_{ij} + f_{jt} + \varepsilon_{ijt} \quad (8.2)$$

α_1 is the constant term. $group$ is a dummy variable which equals 1 if industry j is covered by the pilot ETS and 0 otherwise. β_2 is the DDD estimator measuring the net effect of pilot ETS on industrial carbon emissions. X_{it} denotes a set of controls which might impact the carbon emissions of industry j discussed in more detail below. f_{it}, f_{ij}, f_{jt} and ε_{ijt} denote the region-year fixed effect, the region-industry fixed effect, the industry-year fixed effect and random errors, respectively. Other variables are the same as the above DID model.

To further identify the heterogeneities among different pilot regions and industries, this chapter further introduces two DDD models as follows:

$$E_{ijt} = \alpha_3 + \beta_4 \cdot Treat \cdot time \cdot region + \beta_5 \cdot X_{it} + f_{it} + f_{ij} + f_{jt} + \varepsilon_{ijt} \quad (8.3)$$

$$E_{ijt} = \alpha_4 + \beta_6 \cdot Treat \cdot time \cdot industy + \beta_7 \cdot X_{it} + f_{it} + f_{ij} + f_{jt} + \varepsilon_{ijt} \quad (8.4)$$

In Eq. (8.3), pilot regions are selected as sample. $region$ is a dummy variable which equals 1 if a specific region covered by the pilot ETS is analysed and 0 otherwise. The DDD estimator β_4 is then used to observe the regional heterogeneity in terms of the impact of pilot ETS. Other variables are the same as the Eq. (8.2). In Eq. (8.4), pilot industries are selected as sample. $industry$ is a dummy variable which equals 1 if a specific industry covered by the pilot ETS is analysed and 0 otherwise. The DDD estimator β_6 is then applied to depict the industrial heterogeneity in terms of the impact of pilot ETS. Similarly, other variables are the same as the Eq. (8.2).

8.2.3 Variables and Data

This chapter analyses a compiled provincial-industrial panel dataset covering 26 provincial regions in China's mainland for the 2008–2017 period. Shanghai, Jiangsu, Zhejiang, Sichuan and Tibet are excluded due to data availability. Since 2013, several pilot carbon ETS schemes have been launched, covering selected industries in Beijing, Shanghai, Tianjin, Chongqing, Hubei, Guangdong and Shenzhen. In the practice, the data information of Shenzhen has been statistically covered by the data of Guangdong. Accordingly, the pilot regions in this chapter includes Beijing, Tianjin, Chongqing, Hubei and Guangdong. Industries are identified according to the *Industrial Classification for National Economic Activities* from the National Bureau of Statistics of China.[3] Eight industries in the pilot regions which have been regulated by the pilot ETS are identified as the pilot industries, including papermaking, petrochemical, chemical, building materials, steel, non-ferrous metal, transportation equipment manufacturing and electric power. The relevant data for the complied dataset are from *China Statistical Yearbook* (NBSC 2009–2018c), *China Energy Statistical Yearbook* (NBSC 2009–2018a) and *China Industry Yearbook* (NBSC 2009–2018b).

Industrial carbon emissions are calculated using the energy-based accounting method of IPCC (2015). Specifically, coal, oil and natural gas, which account for more than 84% of the country's energy consumption,[4] are selected as the energy sources for carbon emissions calculation. Direct emissions are included while indirect emissions are not measured in the analysis.

Controls for the provincial-industrial level analysis include output scale, economic scale and industrial performance. Existing literature has shown that those factors can directly or indirectly influence the carbon emissions of industries (Shao et al. 2016; Guo et al. 2020; Trinks et al. 2020; Hou et al. 2021; Tang et al. 2021c; Wu et al. 2021a). Specifically, output scale is measured by an industry's output value (*OV*, 10^8 yuan). Economic scale is measured by an industry's total assets (*ASSET*, 10^8 yuan). Industrial performance is measure by two indicators, including an industry's current assets ratio (*CA*, %) and total profit (*PROFT*, 10^8 yuan). The relevant data for those controls are from *China Statistical Yearbook* (NBSC 2009–2018c) and *China Industry Yearbook* (NBSC 2009–2018b).

8.3 Results and Discussion

8.3.1 Results of the Difference-In-Differences (DID) Analysis

The use of the DID model hinges on the parallel assumption. Accordingly, this chapter needs to evaluate whether or not the identified control group has a similar trend before the treatment as the treatment group. To assess this the analysis compares the annual average carbon emissions for the treatment and control groups (Fig. 8.2). The figure generally depicts that those two groups have similar pre-treatment trends.

---

[3] http://www.stats.gov.cn/tjsj/tjbz/hyflbz/201905/P020190716349644060705.pdf.
[4] https://ourworldindata.org/energy/country/china.

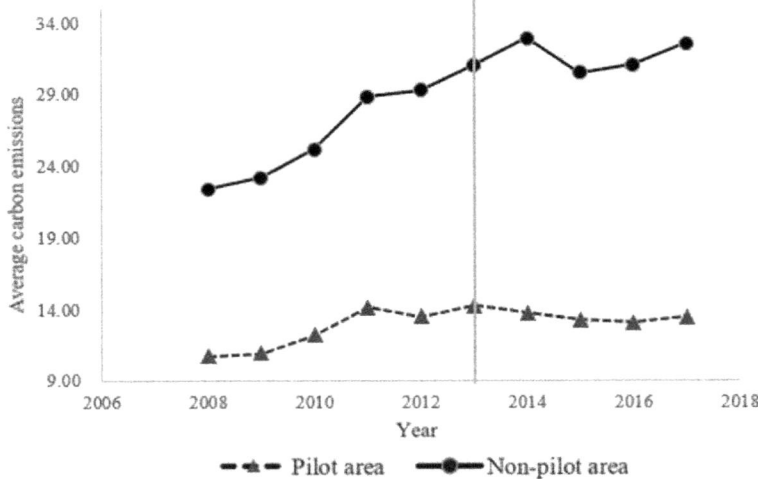

Fig. 8.2 Annual average carbon emissions

After the launch of the pilot ETS, samples in the treatment group tended to decline slowly, whole those in the control group moved in a different direction.

Table 8.2 gives the treatment effect estimates for the DID analysis. Together, the results suggest that carbon emissions of the covered industries in the pilot regions did negatively response to the pilot ETS. After controlling the identified control variables and the three fixed effects, the absolute value of the estimated coefficient of $Treat \cdot time$ is larger than that in Column (1), implying a stronger pilot ETS-induced reduction effect. In terms of the parameter estimates of the control variables, the estimates of output value and total profit are small and significant while those of total assets and current assets ratio are insignificant.

8.3.2 Results of the Difference-In-Difference-In-Differences (DDD) Analysis

Table 8.3 presents the treatment effect estimates for the DDD analysis. All three regressions have considered the region-year, region-industry and industry-year fixed effects. Consistently, the results suggest that the pilot ETS induced carbon emissions reductions in the covered industries in the pilot regions. After controlling the identified control variables, the absolute value of the estimated coefficient of $Treat \cdot time \cdot group$ is slightly larger than that in Column (1). The parameter estimates of the control variables are insignificant.

Furthermore, this chapter compares the estimates of the DID and DDD analyses. Both of them indicate that the launched pilot ETS reduced carbon emissions of

Table 8.2 The average effects of China's pilot ETS on provincial industrial carbon emissions: DID model

Variables	(1)	(2)	(3)	(4)
$Treat \cdot time$	−3.98**	−8.36**	−7.67**	−6.39**
	(1.49)	(3.45)	(2.95)	(2.82)
OV		0.00	0.01**	0.01***
		(0.00)	(0.00)	(0.00)
ASSET		0.01*	0.00	0.00
		(0.01)	(0.00)	(0.00)
CA				0.01
				(0.09)
PROFT				−0.10***
				(0.02)
Cons	−1.08	−14.36*	−14.72**	−15.99**
	(0.67)	(7.93)	(5.52)	(8.01)
Region-fixed effect	Yes	Yes	Yes	Yes
Year-fixed effect	Yes	Yes	Yes	Yes
Industry-fixed effect	No	No	Yes	Yes
R-squared	0.18	0.32	0.52	0.54

Note Standard errors, clustered at the provincial level, are given in parentheses below the estimated parameters; ***, **, and * denote significance levels at the 1%, 5%, and 10%, respectively

Table 8.3 The average effects of China's pilot ETS on provincial industrial carbon emissions: DDD model

Variables	(1)	(2)	(3)
$Treat \cdot time \cdot group$	−3.66**	−3.79**	−3.78**
	(1.47)	(1.53)	(1.55)
OV		−0.00	−0.00
		(0.00)	(0.00)
ASSET		0.00	0.00
		(0.00)	(0.00)
CA			−0.01
			(0.01)
PROFT			−0.00
			(0.00)
Cons	7.55***	7.42***	7.77***
	(0.03)	(0.26)	(0.47)
Region-year fixed effect	Yes	Yes	Yes
Region-industry fixed effect	Yes	Yes	Yes
Industry-year fixed effect	Yes	Yes	Yes
R-squared	0.94	0.95	0.95

Note Standard errors, clustered at the provincial level, are given in parentheses below the estimated parameters; ***, **, and * denote significance levels at the 1%, 5%, and 10%, respectively

covered industries. However, the induced reduction effect from the DDD analysis is smaller than that from the DID analysis. This argues that the use of DID approach may overestimate the reduction effect induced by the pilot ETS since it does not reflect the actual industrial coverage of the scheme.

8.3.3 Results of Robustness Checks

This chapter considers potential threats to robustness. First a placebo test is conducted to evaluate whether or not the estimates are affected by unobservable factors (Gao et al. 2020; Zhou et al. 2020). The main process is to randomly select certain regions as virtual pilot regions in order to compare the differences of the effects between the treatment group and the randomly selected virtual group. Totally, 1,000 random samplings are conducted using the Monte Carlo method. The associated virtual treatment effects are estimated using Eq. (8.2). The t-value distribution of those 1,000 regressions shows that the average coefficient after random samplings is close to 0 compared with the above DDD results. This implies that the DDD estimates are hardly influenced by unobservable factors.

Then a test about concurrent event is conducted. Other regulations launched during the studied period may result in confounding effects and bias the preliminary estimates (Hong et al. 2018; Chen et al. 2019; Wu and Wang 2022). In the analysis two specific regulations that are parallel with the pilot ETS are considered. The first is the water rights trading implemented in Ningxia, Gansu, Inner Mongolia, Henan, Hubei, Jiangxi and Guangdong since 2014.[5] The second is the pilot distributed power generation market trading launched in Henan, Zhejiang, Fujian and Sichuan since 2017.[6] In term of the water rights trading analysis, this chapter excludes the covered regions and reports the results in Columns (1) and (2) of Table 8.4. A similar action has been taken for investigating the possible influence of the pilot distributed power generation market trading and the results are reported in Columns (3) and (4). To sum, the estimates of $Treat \cdot time \cdot group$ in all regressions are consistently negative at the significance level of 5%, arguing that the pilot ETS induced industrial carbon emissions reductions.

8.3.4 Results of Heterogeneity Analysis

Then this chapter explores the heterogeneity of the reduction effect of the pilot ETS on carbon emissions from both the regional and industrial perspectives. First, China's pilot ETS schemes are relatively independently implemented by regional governments in differing ways. For industries regulated, each region selected a different

---

[5] http://szy.mwr.gov.cn/dtxx/201908/t20190806_1352402.html.

[6] http://www.gov.cn/xinwen/2017-11/13/content_5239387.htm.

Table 8.4 Concurrent event test

Variables	Water rights trading		Pilot distributed power generation market trading	
	(1)	(2)	(3)	(4)
Treat · time · group	−3.90**	−4.71**	−4.04**	−4.17**
Cons	(1.50)	(1.76)	(1.58)	(1.65)
	6.98***	5.98***	7.37***	7.60***
	(0.03)	(0.80)	(0.04)	(0.48)
Controls	No	Yes	No	Yes
Region-year fixed effect	Yes	Yes	Yes	Yes
Region-industry fixed effect	Yes	Yes	Yes	Yes
Industry-year fixed effect	Yes	Yes	Yes	Yes
R-squared	0.94	0.94	0.95	0.95

Note Standard errors, clustered at the provincial level, are given in parentheses below the estimated parameters; ***, **, and * denote significance levels at the 1%, 5%, and 10%, respectively

group of industries mainly considering regional industrial and emissions characteristics (Table 8.1). For permits allocation, the approaches used are also different. Most regions use both benchmarking and grandfathering approaches, while Chongqing only employs grandfathering. Those differences may cause regional heterogeneity in terms of the impacts of the pilot ETS.

Table 8.5 shows that the reduction effect of Beijing ETS is significantly larger while that of Guangdong ETS is significantly smaller than the average level. Beijing ETS tends to have higher carbon price than other sampled ETS schemes (Fig. 8.3). Theoretical and empirical evidence has confirmed that higher carbon price, which provides a stronger incentive for emitters, is likely to lead to larger reduction (Lin and Jia 2019; Tang et al. 2019; Wu et al. 2019; Ruhnau et al. 2022). In addition, more stringent regulation in Beijing, the capital of China, may also promote ETS-covered emitters to achieve more reductions in a short term (Chang et al. 2018; Zhang et al. 2019). In terms of the relatively small effect of Guangdong ETS, its relatively low carbon price as a result of excessive permits allocated is a possible cause.

Second, the diverse characteristics of ETS-covered industries may also lead to differing reduction effects of the pilot ETS (Lee and Choi 2018; Shen et al. 2020). Table 8.6 indicates that electric power industry experienced the largest reduction effect induced by the pilot ETS. Electric power industry, which is China's biggest coal user, is the country's top carbon emitter accounting for more than two fifths of the national energy-related carbon emissions (Fig. 8.4).[7] In 2021, this industry released about 4.65 billion tons of CO_2.[8] Therefore, this industry is crucial to achieving the country's net-zero emissions targets and has been identified as the key regulation

---

[7] https://www.iea.org/reports/the-role-of-chinas-ets-in-power-sector-decarbonisation; https://www.nrdc.org/experts/jake-schmidt/chinas-top-industries-can-peak-collective-emissions-2025.

[8] https://www.statista.com/statistics/1300440/power-generation-emissions-china/

Table 8.5 Results of regional heterogeneity analysis

Variables	Regional heterogeneity				
	Beijing	Guangdong	Hubei	Tianjin	Chongqing
$Treat \cdot time \cdot region$	−2.48** (0.81)	2.76** (0.61)	0.28 (0.97)	−1.26 (0.86)	0.58 (0.94)
Cons	4.51*** (0.36)	4.45*** (0.40)	4.62*** (0.43)	4.74*** (0.37)	4.66*** (0.42)
Control variables	Yes	Yes	Yes	Yes	Yes
Region-year fixed effect	Yes	Yes	Yes	Yes	Yes
Region-industry fixed effect	Yes	Yes	Yes	Yes	Yes
Industry-year fixed effect	Yes	Yes	Yes	Yes	Yes
R-squared	0.99	0.98	0.99	0.99	0.99

Note Standard errors, clustered at the provincial level, are given in parentheses below the estimated parameters; ***, **, and * denote significance levels at the 1%, 5%, and 10%, respectively

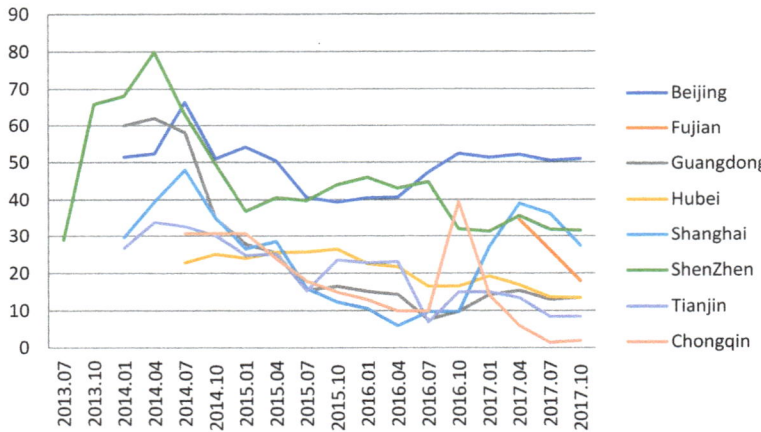

Fig. 8.3 Traded price in first trading day of each quarter from 2013 to 2017 in China's pilot ETS schemes. *Source* Hua and Dong (2019)

area. Accordingly, an increasing number of efforts such as adopting cleaner technologies, updating equipment and improving production process have been taken in this industry as a response to the ETS regulation, thus contributing the large reduction effect.

The coefficients of DDD estimators of papermaking and building materials are significantly positive at the significance level of 5%, implying that the ETS-induced reduction effect in those two industries is smaller than the average level. A possible explanation is that those two industries had been stringently regulated by policies addressing other environmental issues such as water degradation and air pollution before the launch of the pilot ETS. Those regulation policies, i.e., water degradation

Table 8.6 Results of industrial heterogeneity analysis

Variables	Industrial heterogeneity							
	Papermaking	Petrochemical	Chemical	Building materials	Steel	Non-ferrous metal	transportation equipment manufacturing	Electric power
$Treat \cdot time \cdot industy$	6.66**	−3.94	1.20	5.49**	0.55	2.07	−2.06	−10.20**
	(1.91)	(3.17)	(3.65)	(2.31)	(2.76)	(1.79)	(0.47)	(4.80)
Cons	23.78***	23.92***	23.85***	23.73***	23.87***	23.79***	23.82***	23.60***
	(1.98)	(1.98)	(1.98)	(1.97)	(1.98)	(1.95)	(1.99)	(1.89)
Control variables	Yes	Yes	Yes	Yes	Yes	Yes	Yes	Yes
Region-year fixed effect	Yes	Yes	Yes	Yes	Yes	Yes	Yes	Yes
Region-industry fixed effect	Yes	Yes	Yes	Yes	Yes	Yes	Yes	Yes
Industry-year fixed effect	Yes	Yes	Yes	Yes	Yes	Yes	Yes	Yes
R-squared	0.9632	0.9632	0.9631	0.9632	0.9631	0.9631	0.9631	0.9633

Note Standard errors, clustered at the provincial level, are given in parentheses below the estimated parameters; ***, **, and * denote significance levels at the 1%, 5%, and 10%, respectively

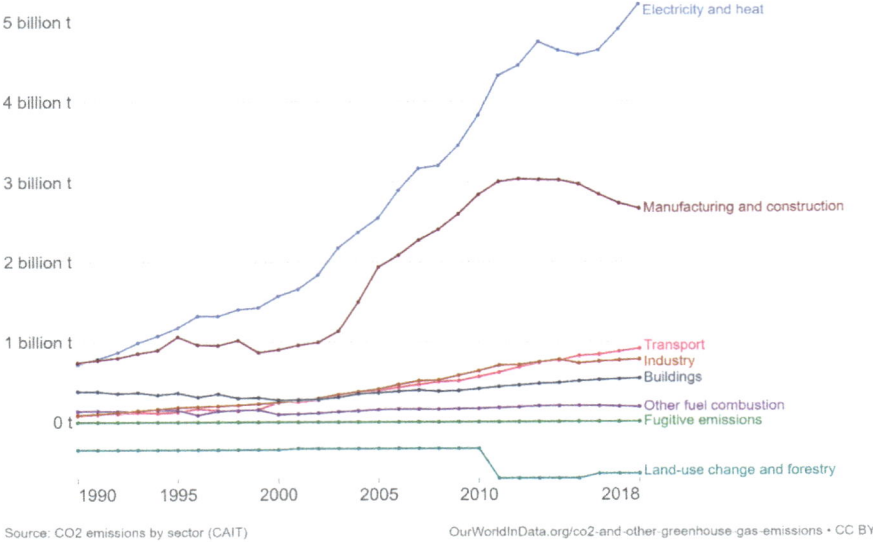

Fig. 8.4 CO_2 emissions by industries, China[9]

and air pollution regulations since 1990s,[10] might generate co-benefits including carbon emissions reductions. Hence, the ETS-induced reduction effect in those two industries is relatively small.

8.3.5 Results of Mechanism Analysis

This chapter also examines potential channels. Specifically, a DID model is proposed to simply verity the inherent channels at the provincial-level as follows:

$$Mediator_{it} = \alpha_5 + \beta_8 \cdot Treat \cdot time + \beta_9 \cdot Z_{it} + f_i + f_t + \varepsilon_{it} \qquad (8.5)$$

where, Z_{it} denotes a set of controls. Other variables are the same as the Eq. (8.1). If β_8 is significant, it proves that the channel exists.

In our case, technological innovation and industrial structure have been identified as possible mediators (Zhou et al. 2019; Zhang et al. 2020; Wu et al. 2021b; Tan et al. 2022; Zhou and Wang 2022). Technological innovation is measured by the ratio of industrial enterprises' internal expenditure on R&D to the value-added of industry. Industrial structure is measured by the ratio of the output value of the second industry to the total output value. In terms of controls, those identified for

---

[9] Figure source: https://ourworldindata.org/grapher/co-emissions-by-sector?country=~CHN.

[10] https://www.mee.gov.cn/gkml/sthjbgw/qt/201310/t20131009_261311.htm.

Table 8.7 Results of mechanism analysis

Variables	Technological innovation				Industrial structure			
	OLS		Fixed effect		OLS		Fixed effect	
	(1)	(2)	(3)	(4)	(5)	(6)	(7)	(8)
Treat · time	0.42**	0.32***	0.42**	0.32***	−0.01	−0.08***	−0.01	−0.08***
	(0.18)	(0.11)	(0.18)	(0.11)	(0.03)	(0.03)	(0.03)	(0.03)
Cons	5.43***	6.13***	5.43***	6.13***	3.16***	2.70***	3.16***	2.70***
	(0.16)	(0.77)	(0.16)	(0.77)	(0.02)	(0.21)	(0.02)	(0.21)
Controls	No	Yes	No	Yes	No	Yes	No	Yes
Region-fixed effect	Yes	Yes	Yes	Yes	Yes	Yes	Yes	Yes
Year-fixed effect	Yes	Yes	Yes	Yes	Yes	Yes	Yes	Yes
R-squared	0.93	0.96	0.67	0.78	0.93	0.95	0.59	0.69

Note Standard errors are given in parentheses below the estimated parameters; ***, **, and * denote significance levels at the 1%, 5%, and 10%, respectively

the previous DID and DDD analyses have been included. Additionally, the proportion of state-owned enterprises has also been controlled due to the consideration of the special connections between state-owned enterprises and governments in China (Tonurist and Karo 2016; Genin et al. 2021; Chen et al. 2022). The relevant data are from *China Statistical Yearbook* (NBSC 2009–2018c) and *China Industry Yearbook* (NBSC 2009–2018b).

The results from OLS regression and fixed effect regression provide evidence that promoting technological innovation and adjusting industrial structure can explain the observed effects on industrial carbon emissions (Table 8.7). The launched pilot ETS could promote technological innovation and industrial structure, which then negatively impact industrial carbon emissions.

8.4 Conclusions

This chapter empirically investigates the impact of China's pilot ETS on provincial industrial carbon emissions. Both DID and DDD approaches are applied to analyse a compiled provincial-industrial level panel dataset covering the 2008–2017 period. The placebo and concurrent evet tests are conducted to check robustness. This chapter also explores the heterogeneity of the identified effect of the pilot ETS from both the regional and industrial perspectives. Additionally, potential channels underlying the relationship between the pilot ETS and industrial carbon emissions are explored.

The results show that the pilot ETS induced carbon emissions reductions in the covered industries in the pilot regions. The reduction effect of Beijing ETS is significantly larger while that of Guangdong ETS is significantly smaller than the average level. Among ETS-covered industries, electric power industry experienced a largest reduction effect induced by the pilot ETS. When examining potential channels, the analysis provides evidence that promoting technological innovation and adjusting industrial structure can explain the observed reduction effect. Overall, China's pilot ETS has contributed to the country's net-zero emissions.

This analysis can be extended by adopting firm-level data to reveal the micro mechanism of China's pilot ETS on carbon emissions. A comparison between the effects of the pilot ETS schemes and the recently launched national carbon market is also needed.

References

Abate, A. G., Riccardi, R., & Ruiz, C. (2021). Contracts in electricity markets under EU ETS: A stochastic programming approach. *Energy Economics*, 99, 105309.

Boyce, J. K. (2018). Carbon pricing: Effectiveness and equity. *Ecological Economics*, 150, 52-61.

Bradley, D., Pantzalis, C., & Yuan, X. (2016). Policy risk, corporate political strategies, and the cost of debt. *Journal of Corporate Finance*, 40, 254-275.

Brauneis, A., Mestel, R., & Palan, S. (2013). Inducing low-carbon investment in the electric power industry through a price floor for emissions trading. *Energy Policy*, 53, 190-204.

Brink, C., Vollebergh, H. R., & van der Werf, E. (2016). Carbon pricing in the EU: Evaluation of different EU ETS reform options. *Energy Policy*, 97, 603-617.

Cai, X., Lu, Y., Wu, M., & Yu, L. (2016). Does environmental regulation drive away inbound foreign direct investment? Evidence from a quasi-natural experiment in China. *Journal of Development Economics*, 123, 73-85.

Caron, J., Rausch, S., & Winchester, N. (2015). Leakage from sub-national climate policy: The case of California's cap–and–trade program. *The Energy Journal*, 36(2), 167-190.

Chan, N. W., & Morrow, J. W. (2019). Unintended consequences of cap-and-trade? Evidence from the Regional Greenhouse Gas Initiative. *Energy Economics*, 80, 411-422.

Chang, K., Ge, F., Zhang, C., & Wang, W. (2018). The dynamic linkage effect between energy and emissions allowances price for regional emissions trading scheme pilots in China. *Renewable and Sustainable Energy Reviews*, 98, 415-425.

Chen, P., Wu, Y., & Zou, L. (2019). Distributive PV trading market in China: A design of multi-agent-based model and its forecast analysis. *Energy*, 185, 423-436.

Chen, Z., Song, P., & Wang, B. (2021). Carbon emissions trading scheme, energy efficiency and rebound effect: Evidence from China's provincial data. *Energy Policy*, 157, 112507.

Chen, X., Chen, G., Lin, M., Tang, K., & Ye, B. (2022). How does anti-corruption affect enterprise green innovation in China's energy-intensive industries?. *Environmental Geochemistry and Health*, 44, 2919–2942.

Choi, Y., Liu, Y., & Lee, H. (2017). The economy impacts of Korean ETS with an emphasis on sectoral coverage based on a CGE approach. *Energy Policy*, 109, 835-844.

Coase, R. (1960). The problem of social cost. *The Journal of Law and Economics*, 3, 1–44.

Convery, F. J. (2009). Origins and development of the EU ETS. *Environmental and Resource Economics*, 43(3), 391-412.

De Perthuis, C., & Trotignon, R. (2014). Governance of CO_2 markets: Lessons from the EU ETS. *Energy Policy*, 75, 100-106.

Diaz-Rainey, I., & Tulloch, D. J. (2018). Carbon pricing and system linking: Lessons from the New Zealand emissions trading scheme. *Energy Economics*, 73, 66-79.

Dong, F., Dai, Y., Zhang, S., Zhang, X., & Long, R. (2019). Can a carbon emission trading scheme generate the Porter effect? Evidence from pilot areas in China. *Science of the Total Environment*, 653, 565-577.

Fageda, X., & Teixidó, J. J. (2022). Pricing carbon in the aviation sector: Evidence from the European emissions trading system. *Journal of Environmental Economics and Management*, 111, 102591.

Fell, H., & Maniloff, P. (2018). Leakage in regional environmental policy: The case of the regional greenhouse gas initiative. *Journal of Environmental Economics and Management*, 87, 1-23.

Gao, Y., Li, M., Xue, J., & Liu, Y. (2020). Evaluation of effectiveness of China's carbon emissions trading scheme in carbon mitigation. *Energy Economics*, 90, 104872.

Genin, A. L., Tan, J., & Song, J. (2021). State governance and technological innovation in emerging economies: State-owned enterprise restructuration and institutional logic dissonance in China's high-speed train sector. *Journal of International Business Studies*, 52(4), 621-645.

Guo, J., Gu, F., Liu, Y., Liang, X., Mo, J., & Fan, Y. (2020). Assessing the impact of ETS trading profit on emission abatements based on firm-level transactions. *Nature Communications*, 11(1), 1-8.

Hintermann, B., Peterson, S., & Rickels, W. (2016). Price and market behavior in phase II of the EU ETS: A review of the literature. *Review of Environmental Economics and Policy*, 10, 108-128.

Hong, B., Zhang, W., Zhou, Y., Chen, J., Xiang, Y., & Mu, Y. (2018). Energy-Internet-oriented microgrid energy management system architecture and its application in China. *Applied Energy*, 228, 2153-2164.

Hou, H., Wang, J., Yuan, M., Liang, S., Liu, T., Wang, H., Bai, H., & Xu, H. (2021). Estimating the mitigation potential of the Chinese service sector using embodied carbon emissions accounting. *Environmental Impact Assessment Review*, 86, 106510.

Hua, Y., & Dong, F. (2019). China's carbon market development and carbon market connection: A literature review. *Energies*, 12(9), 1663.

IPCC. (2015). *Climate Change 2014: Mitigation of Climate Change*. Cambridge University Press, New York.

Kim, W., & Yu, J. (2018). The effect of the penalty system on market prices in the Korea ETS. *Carbon Management*, 9(2), 145-154.

Klenert, D., Mattauch, L., Combet, E., Edenhofer, O., Hepburn, C., Rafaty, R., & Stern, N. (2018). Making carbon pricing work for citizens. *Nature Climate Change*, 8(8), 669-677.

Koch, N., & Mama, H. B. (2019). Does the EU Emissions Trading System induce investment leakage? Evidence from German multinational firms. *Energy Economics*, 81, 479-492.

Lee, H., & Choi, Y. (2018). Heterogeneity and its policy implications in GHG emission performance of manufacturing industries. *Carbon Management*, 9(4), 347-360.

Li, W., & Jia, Z. (2016). The impact of emission trading scheme and the ratio of free quota: A dynamic recursive CGE model in China. *Applied Energy*, 174, 1-14.

Li, Y., Lin, F., & Wang, W. (2021). Environmental regulation and inward foreign direct investment: Evidence from the eleventh Five-Year Plan in China. *Journal of Economic Surveys*. https://doi.org/https://doi.org/10.1111/joes.12439

Lin, B., & Jia, Z. (2018). Impact of quota decline scheme of emission trading in China: A dynamic recursive CGE model. *Energy*, 149, 190-203.

Lin, B., & Jia, Z. (2019). Impacts of carbon price level in carbon emission trading market. *Applied Energy*, 239, 157-170.

Löschel, A., Lutz, B. J., & Managi, S. (2019). The impacts of the EU ETS on efficiency and economic performance: An empirical analyses for German manufacturing firms. *Resource and Energy Economics*, 56, 71-95.

Mildenberger, M., Lachapelle, E., Harrison, K., & Stadelmann-Steffen, I. (2022). Limited impacts of carbon tax rebate programmes on public support for carbon pricing. *Nature Climate Change*, 12, 141-147.

NBSC (National Bureau of Statistics of China) (2009–2018a). *China Energy Statistical Yearbook*. Beijing: China Statistics Press.

NBSC (2011–2017b). *China Industry Statistical Yearbook*. Beijing: China Statistics Press.

NBSC (2011–2017c). *China Statistical Yearbook*. Beijing: China Statistic Press.

Oesingmann, K. (2022). The effect of the European Emissions Trading System (EU ETS) on aviation demand: An empirical comparison with the impact of ticket taxes. *Energy Policy*, 160, 112657.

Pigou, A. C. (1932). *The Economics of Welfare*. 4th ed. London: Macmillan.

Richter, J. L., & Mundaca, L. (2013). Market behavior under the New Zealand ETS. *Carbon Management*, 4(4), 423-438.

Ruhnau, O., Bucksteeg, M., Ritter, D., Schmitz, R., Böttger, D., Koch, M., Pöstges, A., Wiedmann, M. & Hirth, L. (2022). Why electricity market models yield different results: Carbon pricing in a model-comparison experiment. *Renewable and Sustainable Energy Reviews*, 153, 111701.

Shao, S., Liu, J., Geng, Y., Miao, Z., & Yang, Y. (2016). Uncovering driving factors of carbon emissions from China's mining sector. *Applied Energy*, 166, 220-238.

Shawhan, D. L., Taber, J. T., Shi, D., Zimmerman, R. D., Yan, J., Marquet, C. M., Qi, Y., Mao, B., Schuler, R. E., Schulze, W. D. & Tylavsky, D. (2014). Does a detailed model of the electricity grid matter? Estimating the impacts of the Regional Greenhouse Gas Initiative. *Resource and Energy Economics*, 36(1), 191-207.

Shen, B., Dai, F., Price, L., & Lu, H. (2014). California's cap-and-trade programme and insights for China's pilot schemes. *Energy & Environment*, 25(3-4), 551-575.

Shen, J., Tang, P., & Zeng, H. (2020). Does China's carbon emission trading reduce carbon emissions? Evidence from listed firms. *Energy for Sustainable Development*, 59, 120-129.

Skovgaard, J., Ferrari, S. S., & Knaggård, Å. (2019). Mapping and clustering the adoption of carbon pricing policies: What polities price carbon and why?. *Climate Policy*, 19(9), 1173-1185.

Stavins, R. N. (2008). Addressing climate change with a comprehensive US cap-and-trade system. *Oxford Review of Economic Policy*, 298–321.

Stavins, R. N. (2022). The relative merits of carbon pricing instruments: Taxes versus trading. *Review of Environmental Economics and Policy*. https://doi.org/https://doi.org/10.1086/717773

Tan, X., Liu, Y., Cui, J., & Su, B. (2018). Assessment of carbon leakage by channels: An approach combining CGE model and decomposition analysis. *Energy Economics*, 74, 535-545.

Tan, X., Sun, Q., Wang, M., Cheong, T. S., Shum, W. Y., & Huang, J. (2022). Assessing the effects of emissions trading systems on energy consumption and energy mix. *Applied Energy*, 310, 118583.

Tang, K., & Hailu, A. (2020). Smallholder farms' adaptation to the impacts of climate change: Evidence from China's Loess Plateau. *Land Use Policy*, 91, 104353.

Tang, K., & Ma, C. (2022). The cost-effectiveness of agricultural greenhouse gas reduction under diverse carbon policies in China. *China Agricultural Economic Review*. https://doi.org/10.1108/CAER-01-2022-0008

Tang, K., Hailu, A., Kragt, M. E., & Ma, C. (2016a). Marginal abatement costs of greenhouse gas emissions: broadacre farming in the Great Southern Region of Western Australia. *Australian Journal of Agricultural and Resource Economics*, 60, 459-475.

Tang, K., Kragt, M. E., Hailu, A., & Ma, C. (2016b). Carbon farming economics: What have we learned?. *Journal of Environmental Management*, 172, 49-57.

Tang, K., Hailu, A., Kragt, M. E., & Ma, C. (2018). The response of broadacre mixed crop-livestock farmers to agricultural greenhouse gas abatement incentives. *Agricultural Systems*, 160, 11-20.

Tang, K., He, C., Ma, C., & Wang, D. (2019). Does carbon farming provide a cost-effective option to mitigate GHG emissions? Evidence from China. *Australian Journal of Agricultural and Resource Economics*, 63(3), 575-592.

Tang, K., Qiu, Y., & Zhou, D. (2020). Does command-and-control regulation promote green innovation performance? Evidence from China's industrial enterprises. *Science of the Total Environment*, 712, 136362.

Tang, K., Liu, Y., Zhou, D., & Qiu, Y. (2021a). Urban carbon emission intensity under emission trading system in a developing economy: Evidence from 273 Chinese cities. *Environmental Science and Pollution Research*, 28(5), 5168-5179.

Tang, K., Wang, M., & Zhou, D. (2021b). Abatement potential and cost of agricultural greenhouse gases in Australian dryland farming system. *Environmental Science and Pollution Research*, 28(17), 21862-21873.

Tang, K., Xiong, C., Wang, Y., & Zhou, D. (2021c). Carbon emissions performance trend across Chinese cities: Evidence from efficiency and convergence evaluation. *Environmental Science and Pollution Research*, 28(2), 1533-1544.

Tian, Z., Tian, Y., Chen, Y., & Shao, S. (2020). The economic consequences of environmental regulation in China: From a perspective of the environmental protection admonishing talk policy. *Business Strategy and the Environment*, 29(4), 1723-1733.

Tonurist, P., & Karo, E. (2016). State owned enterprises as instruments of innovation policy. *Annals of Public and Cooperative Economics*, 87(4), 623-648.

Trinks, A., Mulder, M., & Scholtens, B. (2020). An efficiency perspective on carbon emissions and financial performance. *Ecological Economics*, 175, 106632.

van den Bergh, J., & Botzen, W. (2020). Low-carbon transition is improbable without carbon pricing. *Proceedings of the National Academy of Sciences*, 117(38), 23219-23220.

Wang, L., & Watanabe, T. (2019). Effects of environmental policy on public risk perceptions of haze in Tianjin City: A difference-in-differences analysis. *Renewable and Sustainable Energy Reviews*, 109, 199-212.

World Bank Group. (2021) State and Trends of Carbon Pricing 2021. Washinton, DC: World Bank. https://openknowledge.worldbank.org/handle/10986/35620

Wu, J., Ma, C., & Tang, K. (2019). The static and dynamic heterogeneity and determinants of marginal abatement cost of CO_2 emissions in Chinese cities. *Energy*, 178, 685-694.

Wu, J., Feng, Z., & Tang, K. (2021a). The dynamics and drivers of environmental performance in Chinese cities: A decomposition analysis. *Environmental Science and Pollution Research*, 28(24), 30626-30641.

Wu, J., Xu, H., & Tang, K. (2021b). Industrial agglomeration, CO_2 emissions and regional development programs: A decomposition analysis based on 286 Chinese cities. *Energy*, 225, 120239.

Wu, L., & Gong, Z. (2021). Can national carbon emission trading policy effectively recover GDP losses? A new linear programming-based three-step estimation approach. *Journal of Cleaner Production*, 287, 125052.

Wu, Q., & Wang, Y. (2022). How does carbon emission price stimulate enterprises' total factor productivity? Insights from China's emission trading scheme pilots. *Energy Economics*, 109, 105990.

Yang, L., Li, Y., & Liu, H. (2021). Did carbon trade improve green production performance? Evidence from China. *Energy Economics*, 96, 105185

Zhang, J., Wang, Z., & Du, X. (2017). Lessons learned from China's regional carbon market pilots. *Economics of Energy & Environmental Policy*, 6(2), 19-38.

Zhang, L., Cao, C., Tang, F., He, J., & Li, D. (2019). Does China's emissions trading system foster corporate green innovation? Evidence from regulating listed companies. *Technology Analysis & Strategic Management*, 31(2), 199-212.

Zhang, Y. J., Shi, W., & Jiang, L. (2020). Does China's carbon emissions trading policy improve the technology innovation of relevant enterprises?. *Business Strategy and the Environment*, 29(3), 872-885.

Zhou, B., Zhang, C., Song, H., & Wang, Q. (2019). How does emission trading reduce China's carbon intensity? An exploration using a decomposition and difference-in-differences approach. *Science of the Total Environment*, 676, 514-523.

Zhou, D., Liang, X., Zhou, Y., & Tang, K. (2020). Does emission trading boost carbon productivity? Evidence from China's pilot emission trading scheme. *International Journal of Environmental Research and Public Health*, 17(15), 5522.

Zhou, F., & Wang, X. (2022). The carbon emissions trading scheme and green technology innovation in China: A new structural economics perspective. *Economic Analysis and Policy*, 74, 365-381.

Chapter 9
Investigating the Impact of Carbon Emission Trading on Industrial Carbon Productivity in China

Di Zhou and Kai Tang

9.1 Introduction

It has been widely thought that carbon pricing is likely to be an essential instrument for efficiently mitigating greenhouse gases (GHG) emissions and achieving the goal of keeping the global temperature increase well below 2 °C (Boyce 2018; Tvinnereim and Mehling 2018; Wilson and Staffell 2018; Skovgaard et al. 2019; Dominioni 2022). Unlike command-and-control regulations commanding who must mitigate GHG emissions where and how, carbon pricing delivers a clear economic signal for emitters via providing a financial incentive to limit emissions. Emitters may choose to reduce their GHG emissions, or could continue emitting and paying for their emissions, thus internalising the external costs of GHG emissions (Pigou 1932; Coase 1960; Stavins 2022). This might explain the widely held belief that the social GHG reduction can be achieved flexibly and cost-effectively with well-functioning carbon pricing (Stavins 2008; van den Bergh and Botzen 2020; Abrell et al. 2022; Tang and Ma 2022).

Carbon pricing can take different forms. The carbon pricing schemes can be classified in emissions trading system (ETS) and carbon taxation schemes according to how they operate. An ETS provides certainty of quantity because it fixes total emissions via tradeable permits and flexible prices, while a carbon tax provides certainty of cost since it directly taxes emissions or explicitly taxes carbon content of commodity (e.g., fuel) (Sumner et al. 2011; Green 2021; Lilliestam et al. 2021;

D. Zhou
School of Mathematics and Statistics, Guangdong University of Foreign Studies, Guangzhou 510006, China

K. Tang (✉)
School of Economics and Trade, Guangdong University of Foreign Studies, Guangzhou 510006, China
e-mail: francistang1988@hotmail.com

Fig. 9.1 Summary map of carbon pricing schemes implemented (Data last updated April 01, 2022). *Source* World Bank (2022)

Dominioni 2022). By incorporating climate change costs into economic decision-making, carbon pricing is likely to encourage switching to carbon–neutral production and consumption, thus underpinning decarbonized development (Tang et al. 2021b; Williams et al. 2021; Dinga and Wen 2022). In addition, carbon pricing generates public revenue for the transition to the carbon–neutral production and consumption, which might be especially helpful considering the enormous public investment needed (Baranzini et al. 2017; Klenert et al. 2018; Steenkamp 2021). In the last two decades, an increasing number of countries and regions have adopted carbon pricing to tackle climate change. There are currently 68 carbon pricing schemes in operation across the globe (Fig. 9.1), covering approximately 23 per cent of global GHG emissions (12 billion tonnes of carbon dioxide equivalent (CO_2e)) (World Bank 2022).

Many have advocated ETS due to its dynamic performance and flexible mechanism (Convery 2009; Christiansen and Wettestad 2003; Stuhlmacher et al. 2019; Tang et al. 2019; Tang and Hailu 2020; Wu et al. 2021). The majority of the established ETS schemes (e.g., EU ETS) are cap-and-trade, but the schemes in China, including the several regional pilot ETS schemes and the recently established national scheme, are baseline-and-credit. Theoretically, both cap-and-trade and baseline-and-credit schemes could be efficient with well-designed frameworks (Buckley et al. 2006, 2008; Woerdman and Nentjes 2019). In practice, however, baseline-and-credit schemes such as China's regional pilot ETS are often criticised as less efficient than cap-and-trade systems in some aspects (e.g., informational function) (Jones and Vossler 2014; Li and Jia 2016; Fan and Todorova 2017; Chang et al. 2019). As the urgent need for response to changing climate, it is necessary to know how well baseline-and-credit schemes perform.

China's regional pilot ETS is the first large-scale carbon ETS implemented in developing economies. Announced in 2011, the pilot ETS has been launched since 2013, covering more than 11% of the country's carbon emissions (Zhang et al. 2017). The established ETS schemes, which are stand-alone regional markets with differing designs, cover selected local major emitting industries. The primary aim of the pilot ETS is to help the largest developing country meet its enhanced climate goals and maintain economic growth simultaneously. In this respect, carbon productivity, which is defined as the amount output per unit of carbon emissions (Ekins et al. 2012; Liu and Zhang 2021), should be a key indicator to evaluate the effectiveness of the pilot ETS since it shows the dynamics between carbon reductions and economic growth (Fan et al. 2021; Jung et al. 2021; Yang et al. 2021a, b).

There is a growing body of literature focusing on the impacts of China's pilot ETS (Mo et al. 2016; Liu et al. 2017; Zhang et al. 2020a, b; Luo et al. 2021; Li et al. 2022a, b). Those studies, adopting either simulation approach (i.e., CGE modelling) or econometric approach based on real-world data, have explored the impacts on diverse socioeconomic indicators. Some have noted that the results of the studies using simulation approaches might be biased since the simulation approaches usually contain multiple assumptions which may conflict with the reality (Yi et al. 2020; Tang et al. 2021a). Moreover, most related econometric studies are conducted at regional level (i.e., provincial- or city-level) or covering all industries, which are inconsistent with the actual coverage of the ETS (Zhang et al. 2020a; Zhu et al. 2020; Li et al. 2022a, b). Overall, few has evaluated the impact of pilot ETS on carbon productivity using real-world data reflecting the actual coverage of the ETS.

This chapter conducts a quasi-natural experiment to empirically investigate the impact of China's pilot ETS on industrial carbon productivity. Analysing provincial industrial-level data in line with the actual ETS coverage, the difference-in-difference-in-difference (DDD) approach is employed. The analysis also assesses the heterogeneous effect of ETS with the consideration of differing pilot regions and industries. In addition, potential channels underlying the relationship between the pilot ETS and industrial carbon productivity are identified. The conclusions may serve as a reference for further defining ETS and better balancing carbon reductions and economic growth in the context of developing economies.

9.2 Methodology

9.2.1 Difference-In-Difference-In-Differences (DDD) Approach

The difference-in-differences (DID) approach is a frequently used tool in evaluating climate policy tools (e.g., Fell and Maniloff 2018; Koch and Mama 2019; Löschel et al. 2019; Tang et al. 2020; Fageda and Teixidó 2022; Wang and Zhang 2022). Though the use of DID approach is able to derive the impact of climate policy (the pilot ETS in this study) to a certain extent, the estimated result might be challenged by the biased issue due to the consideration of varying impact levels on differing industries and regions. In light of this concern, this chapter adopts the DDD approach which modifies the DID approach by considering one more dimension of external variation to better reveal the impact of climate policy (Hsueh 2019; Tang et al. 2021c; Groom et al. 2022; Lu et al. 2022).

In this analysis, ETS-covered industries in the pilot and non-pilot regions are treated as the first treatment and control groups. Other industries in the pilot and non-pilot regions are set as the second treatment and control groups. The use of the second treatment and control groups is expected to eliminate other confounding factors because other industries are not regulated by the ETS schemes.

Specifically, the DDD model is set as:

$$lnCP_{ijt} = \alpha_1 + \beta_0 \cdot Treat \cdot time \cdot group + \beta_1 \cdot X_{it} + f_{it} + f_{jt} + \varepsilon_{ijt} \quad (9.1)$$

where, CP_{ijt} is the carbon productivity of industry j located in region i in year t. α_1 is the constant term. $Treat$ is a dummy variable which equals 1 if industry j is in the pilot region and 0 otherwise. $time$ is a time dummy variable which equals 1 after the establishment of the pilot ETS ($t \geq 2013$) and 0 otherwise. $group$ is a dummy variable which equals 1 if industry j is covered by the pilot ETS and 0 otherwise. β_0 is the DDD estimator indicating the impact degree for covered industries relative to the uncovered ones. X_{it} denotes a set of controls which might impact the carbon productivity of industry j discussed in more detail below. f_{it}, f_{jt} and ε_{ijt} denote the region-year fixed effect, the industry-year fixed effect and random errors, respectively.

9.2.2 Heterogeneous DDD Approach

The established pilot ETS schemes in China are stand-alone regional markets with differing designs in order to reflect local socioeconomic characteristics (e.g., economic level, industrial structure and energy consumption) (Zhang et al. 2017; Wang et al. 2019). Specifically, these pilot schemes are different in key features, including sector coverage choices, permits allocation approaches, enterprises inclusion thresholds, tradable products, compliance rules, and monitoring, reporting and verification methods (Duan et al. 2014; Chen and Xu 2018). Therefore, the impacts of the pilot ETS are likely to vary across piloting regions.

To explore the regional heterogeneity, this chapter selects all piloting regions to form a new sample set and establishes a regional DDD model as follows:

$$lnCP_{ijt} = \alpha_2 + \beta_2 \cdot time \cdot group \cdot location + \beta_3 \cdot X_{it} + f_{it} + f_{jt} + \varepsilon_{ijt}. \quad (9.2)$$

where *location* is a new dummy variable which equals 1 if a specific piloting region is investigated and 0 otherwise. The DDD estimator β_2, our key variable of interest, captures the regional heterogeneity in terms of the impact of pilot ETS on carbon productivity. Other variables are the same as the Eq. (9.1).

It is intuitive that the heterogeneity also exists in the pilot ETS's impact on carbon productivity across covered industries. Some have argued that the impacts of the pilot ETS on technological innovation and enterprises' financial performance present clear industrial heterogeneity (Zhang and Liu 2019; Zhang et al. 2020b). To capture the industrial heterogeneity, this chapter selects all ETS-covered industries to form a new sample set and establishes an industrial DDD model as follows:

$$lnCP_{ijt} = \alpha_3 + \beta_4 \cdot Treat \cdot time \cdot industy + \beta_5 \cdot X_{it} + f_{it} + f_{jt} + \varepsilon_{ijt} \quad (9.3)$$

where *industry* is a new dummy variable which equals 1 if a specific piloting industry is investigated and 0 otherwise. The DDD estimator β_4, our key variable of interest, describes the industrial heterogeneity in terms of the impact on carbon productivity. Other variables are the same as the Eq. (9.1).

9.2.3 Variables and Data

This chapter uses a compiled provincial-industrial panel dataset covering 34 industries in China's mainland from 2008 to 2017. Shanghai and Tibet are not included due to data availability. Since 2013, pilot ETS schemes have been launched in four provincial-level municipalities (Beijing, Shanghai, Tianjin, and Chongqing), two provinces (Hubei and Guangdong), and one sub-provincial municipality (Shenzhen). In the practice, the data information of Shenzhen has been statistically included in the data of Guangdong. Therefore, the piloting regions identified includes Beijing, Tianjin, Hubei, Chongqing and Guangdong. Industries are identified following the *Industrial Classification for National Economic Activities* from the National Bureau of Statistics of China.[1] Pilot industries are those regulated by the pilot ETS in the piloting regions, including papermaking, petrochemical, chemical, building materials, steel, non-ferrous metal, transportation equipment manufacturing and electric power. The relevant data for the used dataset are from *China Statistical Yearbook* (NBSC 2009–2018c), *China Energy Statistical Yearbook* (NBSC 2009–2018a) and *China Industry Yearbook* (NBSC 2009–2018b).

Carbon productivity is defined as the amount industrial gross output per tonne carbon emissions (Ekins et al. 2012; Liu and Zhang 2021). Industrial carbon emissions are estimated according to the energy-based accounting method of IPCC (2015). Coal, oil and natural gas, which account for more than 84% of national energy

---

[1] http://www.stats.gov.cn/tjsj/tjbz/hyflbz/201905/P020190716349644060705.pdf.

consumption,[2] are selected as the energy sources for carbon accounting. Direct emissions are included while indirect emissions are not. Industrial gross output data, adjusted to a 2008 constant price using the Industrial Producer Price Index from the National Bureau of Statistics, are from *China Industry Yearbook* (NBSC 2009–2018b).

The list of industry-specific characterises that are usually included in the analysis include industrial scale and financial performance. This chapter constructs key control variables following the extant literature (Brown et al. 2012; Jaraite-Kažukauske and Di Maria 2016; Xie et al. 2017; Chen et al. 2022). Those control variables may directly or indirectly influence industrial output and/or carbon emissions, thus possibly impacting carbon productivity. Specifically, industrial scale is measured by industrial total assets (*lnTA*, 10^8 yuan, log value) and average number of employees (*lnTL*, 10^4 people, log value). Total assets are measured with a 2008 constant price based on the Fixed Asset Investment Price Index from the National Bureau of Statistics. Financial performance is measured by asset-liability ratio (*ALR*, total liabilities/total assets $\times$ 100%), asset profit ratio (*APR*, total profit/total assets $\times$ 100%), and current assets ratio (*CAR*, total current assets/total assets $\times$ 100%). The relevant data are from *China Industry Yearbook* (NBSC 2009–2018b).

9.3 Results and Discussion

9.3.1 Results of the Overall Impact

Equation (9.1) is employed to deliver the overall impact of pilot ETS on carbon productivity, with the results presented in Table 9.1. Column (1) reports the case without control variables. To explore the robustness of the results, control variables and region-year fixed effect are gradually introduced, as showed in Columns (2) to (4).

As showed in Table 9.1, the coefficients of $Treat \cdot time \cdot group$ persistently remain positive and significant, reaching at least 5% level whether controls are included or not. This implies that the pilot ETS did induce carbon productivity improvement in the covered industries in the piloting regions. In Column (4), the coefficient of $Treat \cdot time \cdot group$ is 0.58, indicating that, compared with the control group, the carbon productivity of the industries covered by the pilot ETS increased by 58 per cent after the launch of the ETS schemes. Considering that the pilot ETS is an exogenous shock to the industries, a conclusion can be made that there exists a significantly positive casual nexus between the pilot ETS and industrial carbon productivity. This argues that the market-based environment regulation could stimulate carbon productivity, which is likely to be consistent with the narrow version of the Porter hypothesis (Jaffe and Palmer 1997; Rassier and Earnhart 2010).

---

[2] https://ourworldindata.org/energy/country/china.

Table 9.1 Impact of China's pilot ETS on industrial carbon productivity

Variables	(1)	(2)	(3)	(4)
$Treat \cdot time \cdot group$	1.23***	0.56**	1.16***	0.58**
	(0.25)	(0.28)	(0.27)	(0.28)
Cons	1.60***	1.62***	0.15	0.82***
	(0.67)	(0.01)	(0.25)	(0.27)
Controls	No	No	Yes	Yes
Industry-year fixed effect	Yes	Yes	Yes	Yes
Region-year fixed effect	No	Yes	No	Yes
R-squared	0.46	0.61	0.50	0.63

Note Standard errors, clustered at the industrial level, are given in parentheses below the estimated parameters; *, ** and *** indicate significance levels at the 10%, 5% and 1% levels, respectively

Theoretically, the market-based regulation such as an ETS means a cost pressure for the regulated producers in the short term. In the long term, however, regulated producers are driven by the market-based regulation to do certain kinds of innovation than do prescriptive regulations, thereby promoting the greening of production (Rassier and Earnhart 2010; Lin et al. 2018). Through compliance regulation and economic compensation, an ETS could induce efficiency improvement and emissions reduction, thereby boosting carbon productivity.

9.3.2   Results of Robustness Checks

To check the robustness, a placebo test is adopted by randomly selecting the piloting regions (Wu and Wang 2022). With the random selection method, it is believed that carbon productivity is not affected by the interaction variable. In practice, the random selections are conducted for 1,000 times. Then, Eq. (9.1) is used to estimate the associated coefficients of the interaction variable. Figure 9.2 shows the t-value distribution of the 1,000 results. The average of the coefficient after 1,000 random samplings is 0.002, close to 0 compared to the above overall impact result and not significant. Therefore, it is believed that the above DDD setting is efficient and the estimated results is reasonable.

A more accurate estimate is required to exclude the potential confounding effect of other concurrent policies. Specifically, the water rights pilots launched in Ningxia, Jiangxi, Inner Mongolia, Gansu, Henan and Hubei since 2014 is considered as an influential policy. Hence, those regions are excluded (Table 9.2). The results indicate that the above outcome has passed the concurrent policy test.

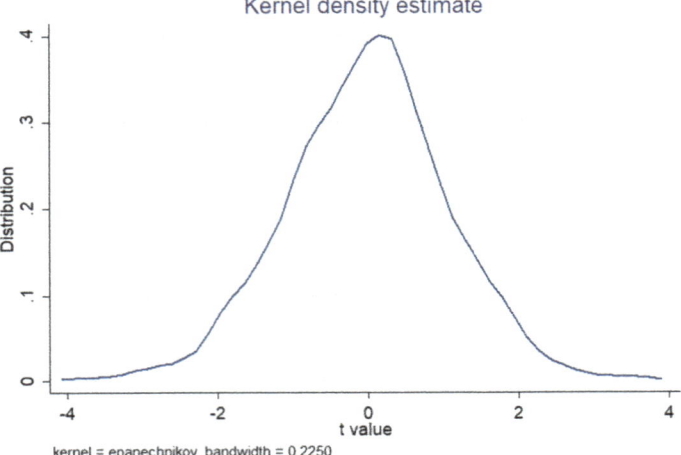

Fig. 9.2 Placebo test

Table 9.2 Concurrent policy test

Variables	Water rights pilots	
	(1)	(2)
$Treat \cdot time \cdot group$	1.13***	0.54**
	(0.183)	(0.24)
Cons	− 0.06*	0.55**
	(0.30)	(0.31)
Controls	Yes	Yes
Industry-year fixed effect	Yes	Yes
Region-year fixed effect	No	Yes
R-squared	0.47	0.61

Note Standard errors, clustered at the industrial level, are given in parentheses below the estimated parameters; *, ** and *** indicate significance levels at the 10%, 5% and 1% levels, respectively

9.3.3 Results of the Heterogeneous Analysis

We first consider the regional heterogeneity of the impact of the pilot ETS on carbon productivity and report the results in Table 9.3. The coefficients of $time \cdot group \cdot location$ for Beijing and Chongqing are significant, while the coefficients for Tianjin, Hubei and Guangdong are not significant.

The Beijing pilot ETS had boosted the carbon productivity levels of covered local industries. It is likely that part of the explanation lies in its relatively higher carbon price (Fig. 9.2). Higher carbon price creates stronger incentives for ETS-covered producers in Beijing, which induce the greening of production and are conducive to

Table 9.3 The heterogeneous effect in regional level

Variables	Regional Heterogeneity				
	Beijing	Tianjin	Hubei	Chongqing	Guangdong
time · group · location	1.07*** (0.37)	0.02 (0.61)	−0.48 (0.51)	−1.29** (0.53)	0.74 (0.46)
Cons	1.06 (0.94)	1.20 (0.96)	1.20 (0.95)	1.14 (0.90)	1.15 (0.92)
Controls	Yes	Yes	Yes	Yes	Yes
Industry-year fixed effect	Yes	Yes	Yes	Yes	Yes
Region-year fixed effect	Yes	Yes	Yes	Yes	Yes
R-squared	0.67	0.66	0.66	0.67	0.67

Note Standard errors, clustered at the industrial level, are given in parentheses below the estimated parameters; *, ** and *** indicate significance levels at the 10%, 5% and 1% levels, respectively

increasing carbon productivity (Jung et al. 2021; Yang et al. 2021a, b; Wu and Wang 2022). In addition, the Beijing pilot ETS has been implemented more stringently in multiple aspects including permits allocation and third-party verification (Duan et al. 2014; Chang et al. 2018; Zhang et al. 2019), which may also contribute to carbon productivity increase.

The coefficients of *time · group · location* for Chongqing is negative at the significance level of 5%, indicating a less efficient ETS in terms of carbon productivity increase. A possible reason is the relatively lower carbon price (Fig. 9.3) due to excessive supply of carbon permits and inactive market performance (Qi and Cheng 2018; Li et al. 2022a, b). Besides, the design and enforcement of Chongqing ETS has been thought to be weak and insufficient (Wang et al. 2020; Yu et al. 2022).

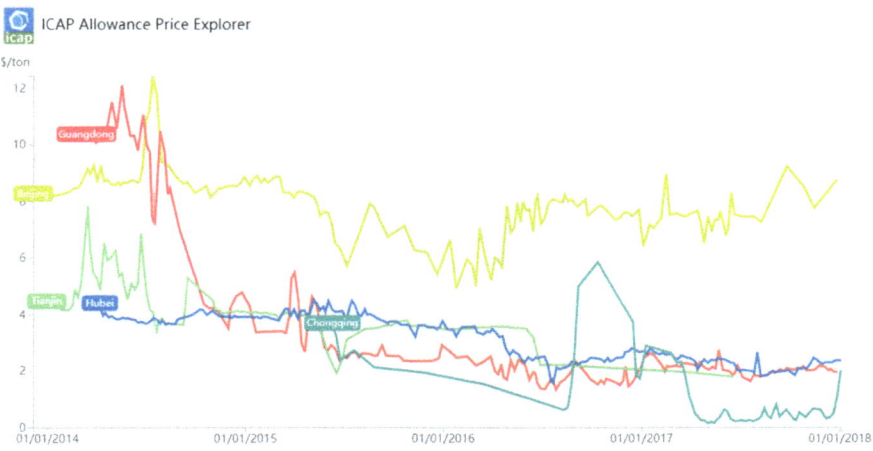

Fig. 9.3 Carbon prices of China's pilot ETS schemes. *Source* ICAP (2022)

We then consider the industrial heterogeneity of the impact of the pilot ETS on carbon productivity. Results in Table 9.4 show that the coefficients of $Treat \cdot time \cdot industy$ for petrochemical and electric power industries are significantly positive, while the coefficients for building materials and transportation industries are significantly negative. The coefficients for papermaking, chemical, steel and non-ferrous metal industries are not significant. Since petrochemical industry experienced the largest carbon productivity promotion effect induced by the pilot ETS, policymakers would need to add this industry to the recently launched national carbon market. In fact, the central government is preparing groundwork to expand the coverage, and is likely to add some of the most polluting industries to the national carbon market in the coming few years.[3]

9.3.4 Results of the Mechanism Analysis

This chapter also examines potential channels. Specifically, Baron and Kenny's stepwise regression (Baron and Kenny 1986), a commonly used mediating test approach, is adopted. Since this study is based on panel data, we use the multi-level mediation method (Krull and MacKinnon 2001) to conduct our mediating analysis.

In our case, technological progress and capital investment have been identified as possible mediators (Bai et al. 2019; Pan et al. 2020; Fan et al. 2021; Sun et al. 2021). Technological progress is measured by the log value of total factor productivity (*TP*) (Zhao and Zhang 2018; Liu et al. 2019). Total factor productivity is estimated using the semiparametric approach proposed by Levinsohn and Petrin (2003). Capital investment (*CI*) is measured by the ratio of industrial fixed assets investment to industrial total output. In terms of controls, those identified for the previous DDD analyses have been included. The relevant data are from *China Statistical Yearbook* (NBSC 2009–2018c) and *China Industry Yearbook* (NBSC 2009–2018b).

The results from stepwise regression show that both the mediating effects of technological progress and capital investment exist, which are partial-mediating (Table 9.5). Therefore, it is evident that promoting technological progress and increasing capital investment can explain the observed effects on industrial carbon productivity. Moreover, the promotion effect of technological progress is stronger than that of capital investment. Overall, the pilot ETS could promote technological progress and capital investment, which then enhance industrial carbon productivity.

---

[3] https://www.bloomberg.com/news/articles/2022-05-31/china-seen-adding-more-sectors-to-carbon-market-in-2024.

Table 9.4 The heterogeneous effect at industrial level

Variables	Industrial heterogeneity							
	Papermaking	Petrochemical	Chemical	Building materials	Steel	Non-ferrous metal	Transportation	Electric power
$Treat \cdot time \cdot industy$	−0.11	1.39***	0.04	−0.51*	−0.35	−0.34	−0.92**	0.74**
	(0.29)	(0.19)	(0.26)	(0.25)	(0.30)	(0.26)	(0.28)	(0.30)
Cons	0.41	0.31	0.41	0.42	0.42	0.45	0.34	0.43
	(0.47)	(0.51)	(0.46)	(0.47)	(0.46)	(0.46)	(0.49)	(0.45)
Controls	Yes	Yes	Yes	Yes	Yes	Yes	Yes	Yes
Industry-year fixed effect	Yes	Yes	Yes	Yes	Yes	Yes	Yes	Yes
Region-year fixed effect	Yes	Yes	Yes	Yes	Yes	Yes	Yes	Yes
R-squared	0.70	0.70	0.69	0.70	0.70	0.69	0.70	0.70

Note Standard errors, clustered at the industrial level, are given in parentheses below the estimated parameters; *, ** and *** indicate significance levels at the 10%, 5% and 1% levels, respectively

Table 9.5 Median effect test: stepwise regression

Variables	Technological progress				Capital investment			
	(1)	(2)	(3)	(4)	(5)	(6)	(7)	(8)
$Treat \cdot time \cdot group$	0.58** (0.28)			0.48* (0.28)	0.58** (0.28)			0.57** (0.28)
TP		0.63*** (0.13)	0.21* (0.12)	0.22** (0.05)				
CI						0.05 * (0.03)	0.07 * (0.12)	0.01*** (0.05)
Cons	0.82*** (0.27)	−1.28*** (0.27)	−1.03*** (0.25)	1.31*** (0.28)	0.82*** (0.27)	1.06*** (0.36)	−1.62 (1.84)	1.31*** (0.00)
Controls	Yes	Yes	Yes	Yes	Yes	Yes	Yes	Yes
Industry-year fixed effect	Yes	Yes	Yes	Yes	Yes	Yes	Yes	Yes
Region-year fixed effect	Yes	No	Yes	Yes	Yes	No	No	Yes
Region fixed effect	No	No	No	No	No	No	Yes	No
R-squared	0.63	0.22	0.83	0.64	0.63	0.09	0.09	0.63

Note Standard errors, clustered at the industrial level, are given in parentheses below the estimated parameters; *, ** and *** indicate significance levels at the 10%, 5% and 1% levels, respectively

9.4 Conclusions

This chapter conducts a quasi-natural experiment to empirically investigate the impact of China's pilot ETS on industrial carbon productivity. The DDD approach is adopted to analyse a provincial industrial-level dataset in line with the actual pilot ETS coverage. The dataset contains the information of economic variables and carbon emissions from 2008 to 2017. The placebo and concurrent policy tests are employed to check the robustness of the estimated results. The heterogeneity of the identified effect of the pilot ETS is explored from both the regional and industrial perspectives. In addition, potential channels underlying the relationship between the pilot ETS and industrial carbon productivity are explored.

The baseline regression confirms that the pilot ETS induced carbon productivity improvement. There exists a significantly positive casual nexus between the pilot ETS and industrial carbon productivity. This outcome remains robust through the placebo and concurrent policy tests. Moreover, the promotion effect of Beijing ETS is significantly larger while that of Chongqing ETS is significantly smaller than the average level. Among ETS-covered industries, petrochemical industry experienced the largest promotion effect induced by the pilot ETS. When examining potential channels, the analysis provides evidence that enhancing technological progress and increasing capital investment can explain the observed promotion effect. In general, China's pilot ETS has contributed to the country's carbon productivity improvement.

This analysis can be extended by making a comparison between the effects of the pilot ETS and the recently launched national carbon market. Also, a comparison between China's baseline-and-credit schemes and other countries' cap-and-trade schemes will be useful.

References

Abrell, J., Kosch, M., & Rausch, S. (2022). How effective is carbon pricing? A machine learning approach to policy evaluation. *Journal of Environmental Economics and Management*, 112, 102589.

Bai, C., Du, K., Yu, Y., & Feng, C. (2019). Understanding the trend of total factor carbon productivity in the world: Insights from convergence analysis. *Energy Economics*, 81, 698-708.

Baranzini, A., Van den Bergh, J. C., Carattini, S., Howarth, R. B., Padilla, E., & Roca, J. (2017). Carbon pricing in climate policy: Seven reasons, complementary instruments, and political economy considerations. *Wiley Interdisciplinary Reviews: Climate Change*, 8(4), e462.

Baron, R. M., & Kenny, D. A. (1986). The moderator–mediator variable distinction in social psychological research: Conceptual, strategic, and statistical considerations. *Journal of Personality and Social Psychology*, 51(6), 1173.

Boyce, J. K. (2018). Carbon pricing: Effectiveness and equity. *Ecological Economics*, 150, 52-61.

Brown, J. R., Martinsson, G., & Petersen, B. C. (2012). Do financing constraints matter for R&D?. *European Economic Review*, 56(8), 1512-1529.

Buckley, N. J., Mestelman, S., & Muller, R. A. (2006). Implications of alternative emission trading plans: Experimental evidence. *Pacific Economic Review*, 11(2), 149-166.

Buckley, N. J., Mestelman, S., & Muller, R. A. (2008). Baseline-and-credit emission permit trading. In T. L. Cherry, S. Kroll, & J. F. Shogren (Eds.), *Environmental Economics, Experimental Methods* (pp. 9–28). Routledge Explorations in Environmental Economics. London and New York: Taylor and Francis, Routledge.

Chang, K., Ge, F., Zhang, C., & Wang, W. (2018). The dynamic linkage effect between energy and emissions allowances price for regional emissions trading scheme pilots in China. *Renewable and Sustainable Energy Reviews*, 98, 415-425.

Chang, K., Ye, Z., & Wang, W. (2019). Volatility spillover effect and dynamic correlation between regional emissions allowances and fossil energy markets: New evidence from China's emissions trading scheme pilots. *Energy*, 185, 1314-1324.

Chen, X., & Xu, J. (2018). Carbon trading scheme in the People's Republic of China: Evaluating the performance of seven pilot projects. *Asian Development Review*, 35(2), 131-152.

Chen, X., Chen, G., Lin, M., Tang, K., & Ye, B. (2022). How does anti-corruption affect enterprise green innovation in China's energy-intensive industries?. *Environmental Geochemistry and Health*, 44, 2919-2942.

Christiansen, A. C., & Wettestad, J. (2003). The EU as a frontrunner on greenhouse gas emissions trading: How did it happen and will the EU succeed?. *Climate Policy*, 3(1), 3-18.

Coase, R. (1960). The problem of social cost. *The Journal of Law and Economics*, 3, 1–44.

Convery, F. J. (2009). Origins and development of the EU ETS. *Environmental and Resource Economics*, 43(3), 391-412.

Dinga, C. D., & Wen, Z. (2022). China's green deal: Can China's cement industry achieve carbon neutral emissions by 2060?. *Renewable and Sustainable Energy Reviews*, 155, 111931.

Dominioni, G. (2022). Pricing carbon effectively: A pathway for higher climate change ambition. *Climate Policy*. https://doi.org/10.1080/14693062.2022.2042177

Duan, M., Pang, T., & Zhang, X. (2014). Review of carbon emissions trading pilots in China. *Energy & Environment*, 25(3-4), 527-549.

Ekins, P., Pollitt, H., Summerton, P., & Chewpreecha, U. (2012). Increasing carbon and material productivity through environmental tax reform. *Energy Policy*, 42, 365-376.

Fageda, X., & Teixidó, J. J. (2022). Pricing carbon in the aviation sector: Evidence from the European emissions trading system. *Journal of Environmental Economics and Management*, 111, 102591.

Fan, J. H., & Todorova, N. (2017). Dynamics of China's carbon prices in the pilot trading phase. *Applied Energy*, 208, 1452-1467.

Fan, L. W., You, J., & Zhou, P. (2021). How does technological progress promote carbon productivity? Evidence from Chinese manufacturing industries. *Journal of Environmental Management*, 277, 111325.

Fell, H., & Maniloff, P. (2018). Leakage in regional environmental policy: The case of the regional greenhouse gas initiative. *Journal of Environmental Economics and Management*, 87, 1-23.

Green, J. F. (2021). Does carbon pricing reduce emissions? A review of ex-post analyses. *Environmental Research Letters*, 16(4), 043004.

Groom, B., Palmer, C., & Sileci, L. (2022). Carbon emissions reductions from Indonesia's moratorium on forest concessions are cost-effective yet contribute little to Paris pledges. *Proceedings of the National Academy of Sciences*, 119(5), e2102613119.

Hsueh, L. (2019). Voluntary climate action and credible regulatory threat: Evidence from the carbon disclosure project. *Journal of Regulatory Economics*, 56(2), 188-225.

ICAP (2022). ICAP Allowance Price Explorer. International Carbon Action Partnership. https://icapcarbonaction.com/en/ets-prices

IPCC. (2015). *Climate Change 2014: Mitigation of Climate Change*. Cambridge University Press, New York.

Jaffe, A. B., & Palmer, K. (1997). Environmental regulation and innovation: A panel data study. *Review of Economics and Statistics*, 79(4), 610-619.

Jaraite-Kažukauske, J., & Di Maria, C. (2016). Did the EU ETS make a difference? An empirical assessment using Lithuanian firm-level data. *The Energy Journal*, 37(1), 1-23.

Jones, L. R., & Vossler, C. A. (2014). Experimental tests of water quality trading markets. *Journal of Environmental Economics and Management*, 68(3), 449-462.

Jung, H., Song, S., Ahn, Y. H., Hwang, H., & Song, C. K. (2021). Effects of emission trading schemes on corporate carbon productivity and implications for firm-level responses. *Scientific Reports*, 11(1), 11679.

Klenert, D., Mattauch, L., Combet, E., Edenhofer, O., Hepburn, C., Rafaty, R., & Stern, N. (2018). Making carbon pricing work for citizens. *Nature Climate Change*, 8(8), 669-677.

Koch, N., & Mama, H. B. (2019). Does the EU Emissions Trading System induce investment leakage? Evidence from German multinational firms. *Energy Economics*, 81, 479-492.

Krull, J. L., & MacKinnon, D. P. (2001). Multilevel modeling of individual and group level mediated effects. *Multivariate Behavioral Research*, 36(2), 249-277.

Levinsohn, J., & Petrin, A. (2003). Estimating production functions using inputs to control for unobservables. *The Review of Economic Studies*, 70(2), 317-341.

Li, C., Qi, Y., Liu, S., & Wang, X. (2022). Do carbon ETS pilots improve cities' green total factor productivity? Evidence from a quasi-natural experiment in China. *Energy Economics*, 108, 105931.

Li, K., Qi, S. Z., Yan, Y. X., & Zhang, X. L. (2022). China's ETS pilots: Program design, industry risk, and long-term investment. *Advances in Climate Change Research*, 13(1), 82-96.

Li, W., & Jia, Z. (2016). The impact of emission trading scheme and the ratio of free quota: A dynamic recursive CGE model in China. *Applied Energy*, 174, 1-14.

Lilliestam, J., Patt, A., & Bersalli, G. (2021). The effect of carbon pricing on technological change for full energy decarbonization: A review of empirical ex-post evidence. *Wiley Interdisciplinary Reviews: Climate Change*, 12(1), e681.

Lin, S., Wang, B., Wu, W., & Qi, S. (2018). The potential influence of the carbon market on clean technology innovation in China. *Climate Policy*, 18, 71-89.

Liu, H., Wu, J., & Chu, J. (2019). Environmental efficiency and technological progress of transportation industry-based on large scale data. *Technological Forecasting and Social Change*, 144, 475-482.

Liu, X., & Zhang, X. (2021). Industrial agglomeration, technological innovation and carbon productivity: Evidence from China. *Resources, Conservation and Recycling*, 166, 105330.

Liu, Y., Tan, X. J., Yu, Y., & Qi, S. Z. (2017). Assessment of impacts of Hubei Pilot emission trading schemes in China–A CGE-analysis using TermCO2 model. *Applied Energy*, 189, 762-769.

Löschel, A., Lutz, B. J., & Managi, S. (2019). The impacts of the EU ETS on efficiency and economic performance: An empirical analyses for German manufacturing firms. *Resource and Energy Economics*, 56, 71-95.

Lu, S., Li, S., Zhou, W., & Yang, W. (2022). Network herding of energy funds in the post-Carbon-Peak Policy era: Does it benefit profitability and stability?. *Energy Economics*, 109, 105948.

Luo, Y., Li, X., Qi, X., & Zhao, D. (2021). The impact of emission trading schemes on firm competitiveness: Evidence of the mediating effects of firm behaviors from the Guangdong ETS. *Journal of Environmental Management*, 290, 112633.

Mo, J. L., Agnolucci, P., Jiang, M. R., & Fan, Y. (2016). The impact of Chinese carbon emission trading scheme (ETS) on low carbon energy (LCE) investment. *Energy Policy*, 89, 271-283.

NBSC (National Bureau of Statistics of China) (2009–2018a). *China Energy Statistical Yearbook*. Beijing: China Statistics Press.

NBSC (2011–2017b). *China Industry Statistical Yearbook*. Beijing: China Statistics Press.

NBSC (2011–2017c). *China Statistical Yearbook*. Beijing: China Statistic Press.

Pan, X., Li, M., Wang, M., Chu, J., & Bo, H. (2020). The effects of outward foreign direct investment and reverse technology spillover on China's carbon productivity. *Energy Policy*, 145, 111730.

Pigou, A. C. (1932). *The Economics of Welfare*. 4th ed. London: Macmillan.

Qi, S., & Cheng, S. (2018). China's national emissions trading scheme: Integrating cap, coverage and allocation. *Climate Policy*, 18, 45-59.

Rassier, D. G., & Earnhart, D. (2010). Does the porter hypothesis explain expected future financial performance? The effect of clean water regulation on chemical manufacturing firms. *Environmental and Resource Economics*, 45(3), 353-377.

Skovgaard, J., Ferrari, S. S., & Knaggård, Å. (2019). Mapping and clustering the adoption of carbon pricing policies: What polities price carbon and why?. *Climate Policy*, 19(9), 1173-1185.

Stavins, R. N. (2008). Addressing climate change with a comprehensive US cap-and-trade system. *Oxford Review of Economic Policy*, 298–321.

Stavins, R. N. (2022). The relative merits of carbon pricing instruments: Taxes versus trading. *Review of Environmental Economics and Policy*. https://doi.org/https://doi.org/10.1086/717773

Steenkamp, L. A. (2021). A classification framework for carbon tax revenue use. *Climate Policy*, 21(7), 897-911.

Stuhlmacher, M., Patnaik, S., Streletskiy, D., & Taylor, K. (2019). Cap-and-trade and emissions clustering: A spatial-temporal analysis of the European Union Emissions Trading Scheme. *Journal of Environmental Management*, 249, 109352.

Sumner, J., Bird, L., & Dobos, H. (2011). Carbon taxes: A review of experience and policy design considerations. *Climate Policy*, 11(2), 922-943.

Sun, L. X., Xia, Y. S., & Feng, C. (2021). Income gap and global carbon productivity inequality: A meta-frontier data envelopment analysis. *Sustainable Production and Consumption*, 26, 548-557.

Tang, K., & Hailu, A. (2020). Smallholder farms' adaptation to the impacts of climate change: Evidence from China's Loess Plateau. *Land Use Policy*, 91, 104353.

Tang, K., He, C., Ma, C., & Wang, D. (2019). Does carbon farming provide a cost-effective option to mitigate GHG emissions? Evidence from China. *Australian Journal of Agricultural and Resource Economics*, 63(3), 575-592.

Tang, K., Liu, Y., Zhou, D., & Qiu, Y. (2021a). Urban carbon emission intensity under emission trading system in a developing economy: Evidence from 273 Chinese cities. *Environmental Science and Pollution Research*, 28(5), 5168-5179.

Tang, K., & Ma, C. (2022). The cost-effectiveness of agricultural greenhouse gas reduction under diverse carbon policies in China. *China Agricultural Economic Review*. https://doi.org/10.1108/CAER-01-2022-0008

Tang, K., Qiu, Y., & Zhou, D. (2020). Does command-and-control regulation promote green innovation performance? Evidence from China's industrial enterprises. *Science of the Total Environment*, 712, 136362.

Tang, K., Wang, M., & Zhou, D. (2021b). Abatement potential and cost of agricultural greenhouse gases in Australian dryland farming system. *Environmental Science and Pollution Research*, 28(17), 21862-21873.

Tang, K., Zhou, Y., Liang, X., & Zhou, D. (2021c). The effectiveness and heterogeneity of carbon emissions trading scheme in China. *Environmental Science and Pollution Research*, 28(14), 17306-17318.

Tvinnereim, E., & Mehling, M. (2018). Carbon pricing and deep decarbonisation. *Energy Policy*, 121, 185-189.

van den Bergh, J., & Botzen, W. (2020). Low-carbon transition is improbable without carbon pricing. *Proceedings of the National Academy of Sciences*, 117(38), 23219-23220.

Wang, B., Boute, A., & Tan, X. (2020). Price stabilization mechanisms in China's pilot emissions trading schemes: Design and performance. *Climate Policy*, 20(1), 46-59.

Wang, P., Liu, L., Tan, X., & Liu, Z. (2019). Key challenges for China's carbon emissions trading program. *Wiley Interdisciplinary Reviews: Climate Change*, 10(5), e599.

Wang, W., & Zhang, Y. J. (2022). Does China's carbon emissions trading scheme affect the market power of high-carbon enterprises?. *Energy Economics*, 108, 105906.

Williams, J. H., Jones, R. A., Haley, B., Kwok, G., Hargreaves, J., Farbes, J., & Torn, M. S. (2021). Carbon-neutral pathways for the United States. *AGU Advances*, 2(1), e2020AV000284.

Wilson, I. A., & Staffell, I. (2018). Rapid fuel switching from coal to natural gas through effective carbon pricing. *Nature Energy*, 3(5), 365-372.

Woerdman, E., & Nentjes, A. (2019). Emissions trading hybrids: The case of the EU ETS. *Review of Law & Economics*, 15(1).

World Bank (2022). Carbon Pricing Dashboard. https://carbonpricingdashboard.worldbank.org/map_data (accessed 9 June 2022)

Wu, J., Feng, Z., & Tang, K. (2021). The dynamics and drivers of environmental performance in Chinese cities: a decomposition analysis. *Environmental Science and Pollution Research*, 28(24), 30626-30641.

Wu, Q., & Wang, Y. (2022). How does carbon emission price stimulate enterprises' total factor productivity? Insights from China's emission trading scheme pilots. *Energy Economics*, 109, 105990.

Xie, R. H., Yuan, Y. J., & Huang, J. J. (2017). Different types of environmental regulations and heterogeneous influence on "green" productivity: Evidence from China. *Ecological Economics*, 132, 104-112.

Yang, H., Lu, Z., Shi, X., Mensah, I. A., Luo, Y., & Chen, W. (2021). Multi-region and multi-sector comparisons and analysis of industrial carbon productivity in China. *Journal of Cleaner Production*, 279, 123623.

Yang, L., Li, Y., & Liu, H. (2021b). Did carbon trade improve green production performance? Evidence from China. *Energy Economics*, 96, 105185

Yi, L., Bai, N., Yang, L., Li, Z., & Wang, F. (2020). Evaluation on the effectiveness of China's pilot carbon market policy. *Journal of Cleaner Production*, 246, 119039.

Yu, P., Hao, R., Cai, Z., Sun, Y., & Zhang, X. (2022). Does emission trading system achieve the win-win of carbon emission reduction and financial performance improvement? Evidence from Chinese A-share listed firms in industrial sector. *Journal of Cleaner Production*, 333, 130121.

Zhang, J., Wang, Z., & Du, X. (2017). Lessons learned from China's regional carbon market pilots. *Economics of Energy & Environmental Policy*, 6(2), 19-38.

Zhang, L., Cao, C., Tang, F., He, J., & Li, D. (2019). Does China's emissions trading system foster corporate green innovation? Evidence from regulating listed companies. *Technology Analysis & Strategic Management*, 31(2), 199-212.

Zhang, G., & Zhang, N. (2020). The effect of China's pilot carbon emissions trading schemes on poverty alleviation: A quasi-natural experiment approach. *Journal of Environmental Management*, 271, 110973.

Zhang, Y. J., Liang, T., Jin, Y. L., & Shen, B. (2020a). The impact of carbon trading on economic output and carbon emissions reduction in China's industrial sectors. *Applied Energy*, 260, 114290.

Zhang, Y. J., Shi, W., & Jiang, L. (2020b). Does China's carbon emissions trading policy improve the technology innovation of relevant enterprises?. *Business Strategy and the Environment*, 29(3), 872-885.

Zhang, Y. J., & Liu, J. Y. (2019). Does carbon emissions trading affect the financial performance of high energy-consuming firms in China?. *Natural Hazards*, 95(1), 91-111.

Zhao, X., & Zhang, Y. (2018). Technological progress and industrial performance: A case study of solar photovoltaic industry. *Renewable and Sustainable Energy Reviews*, 81, 929-936.

Zhu, B., Zhang, M., Huang, L., Wang, P., Su, B., & Wei, Y. M. (2020). Exploring the effect of carbon trading mechanism on China's green development efficiency: A novel integrated approach. *Energy Economics*, 85, 104601

Chapter 10
Carbon Sequestration and Greenhouse Gas Emissions Reductions in Agriculture: Strategies and Their Economic Feasibility

Kai Tang

10.1 Introduction

As one of the key sources of greenhouse gas (GHG) emissions, agriculture contributes large amounts of carbon dioxide (CO_2), methane (CH_4) and nitrous oxide (N_2O) through, i.e., emissions from cropping or livestock (Wollenberg et al. 2016; Tang et al. 2016b, 2018; Tongwane and Moeletsi 2018; Crippa et al. 2021). Globally, agricultural activities contribute about 30 per cent of total anthropogenic GHG emissions, 50% of CH_4 emissions and 75% of N_2O emissions (Saunois et al. 2020; Gütschow et al. 2021; FAO 2022).[1] In 2019, global GHG emissions from crop and livestock production were 7.2 billion tonnes carbon dioxide equivalent (CO_2e) (Tubiello et al. 2022) (Fig. 10.1). Agricultural GHG emissions contribute an average of 35% of emissions in developing economies and 12% in developed economies (Richards et al. 2015). In China, agriculture accounts for approximately 15% of total GHG emissions, 90% of N_2O emissions, and 60% of CH_4 emissions (Tang et al. 2016a, 2019; Tang and Ma 2022). In 2019, GHG emissions from crop and livestock production in China were about 0.8 billion tonnes CO_2e (Fig. 10.2).

A growing body of literature has shown that agriculture exerts a significant role in achieving global net-zero emissions (Bennetzen et al. 2016; Lamb et al. 2016; Davis et al. 2018; Tang et al. 2018; Laborde et al. 2021; Northrup et al. 2021; Fujimori et al. 2022). Agroecosystems can lock up carbon on land and in the grown plants, thus having a large carbon sink capacity. Biophysical carbon sequestration in agricultural systems takes atmospheric CO_2 and converts it into carbon stored in agricultural soils (Fig. 10.3). The global organic carbon pool in the upper 1 m of soils contains 1,325

---

[1] https://www.iaea.org/topics/greenhouse-gas-reduction.

K. Tang (✉)
School of Economics and Trade, Guangdong University of Foreign Studies, Guangzhou 510006, China
e-mail: francistang1988@hotmail.com

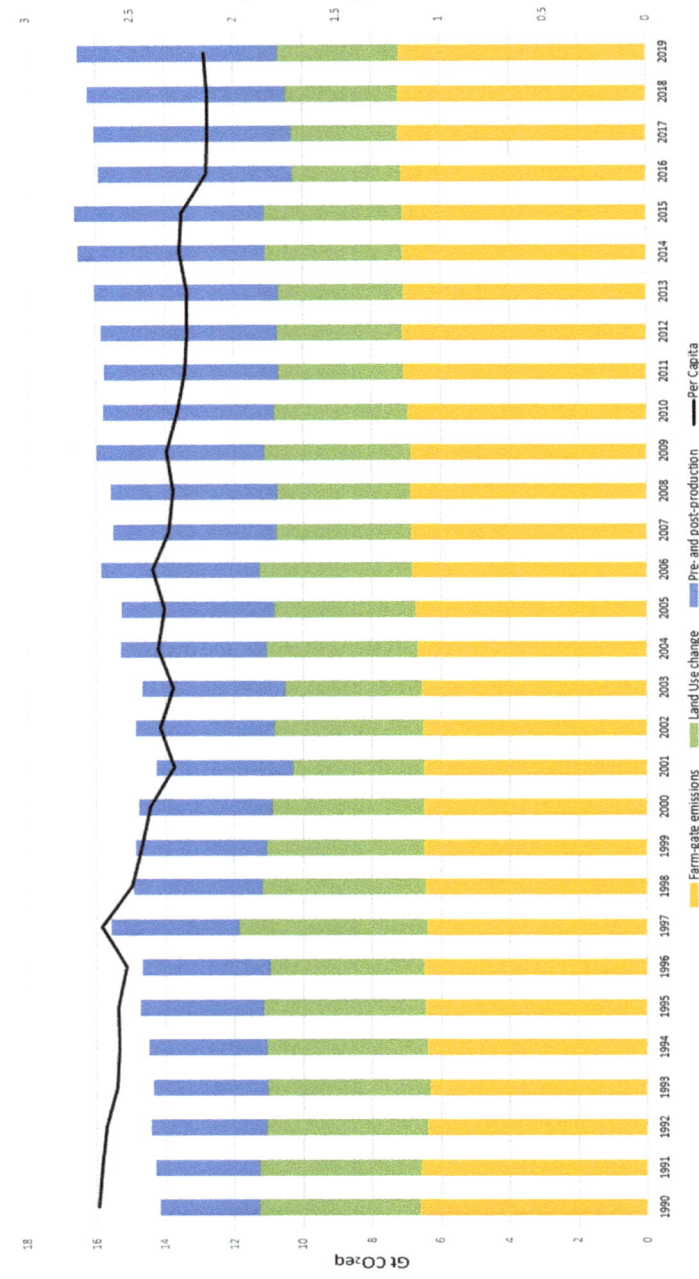

Fig. 10.1 Global total GHG emissions from agri-food systems, 1990–2019. *Source* Tubiello et al. (2022)

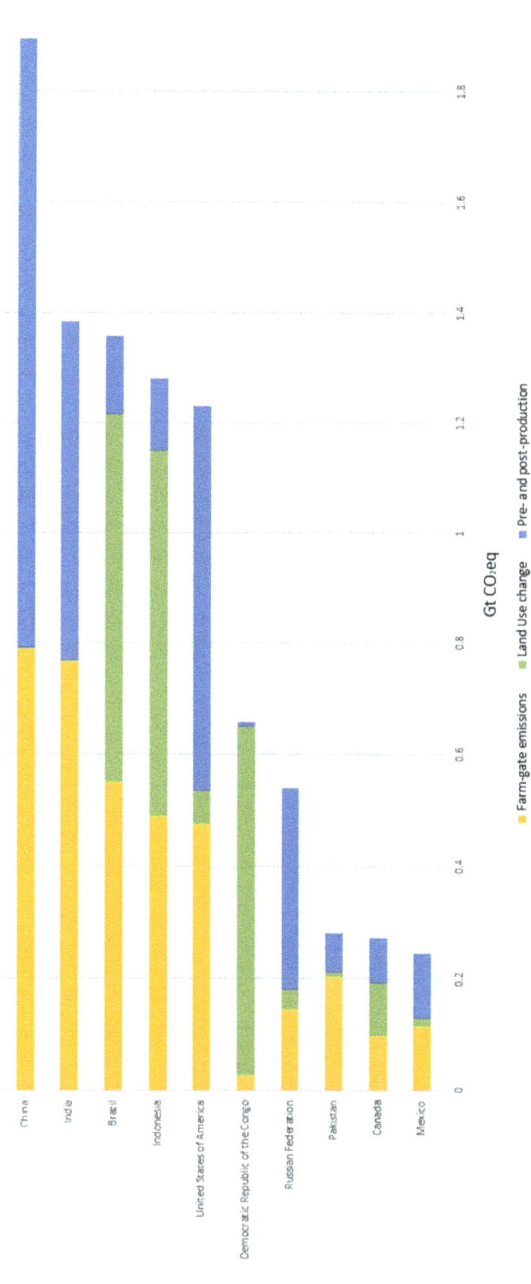

Fig. 10.2 Total GHG emission from agri-food systems by countries, 2019. *Source* Tubiello et al. (2022)

billion tonnes carbon, which is more than three times the amount of carbon in Earth's atmosphere (Köchy et al. 2015). The annual rate of carbon sequestration ranges from 0.1 to 1 tonne carbon per hectare for soil organic carbon, and the rate is relatively high in heavy or clayey soils with deep solum of humid and cool climates (Lal et al. 2015). Researchers have argued that a 4 per mille increase in soil carbon might compensate the global anthropogenic GHG emissions each year (Minasny et al. 2017). A shift in agricultural practices that promote crop growth and increase productivity is likely to sequestrate more soil carbon (Chowdhury et al. 2021). In addition, agricultural GHG emissions can be reduced through changing cropping and livestock production activities (Thamo et al. 2013; Daigneault et al. 2018; Tang et al. 2018, 2019; Tang and Hailu 2020). Adopting improved agricultural technologies and practices has the potential to achieve approximately 20% of agriculture's required emissions reduction by the middle of this century (Ahmed et al. 2020). Technically, the annual agricultural GHG mitigation potential is about 5,500–6, 000 million tonnes CO_2e (Smith et al. 2008). As such, agriculture has the great potential to make a substantial contribution to net-zero emissions progress.

Carbon sequestration and GHG emissions reductions in agriculture is not well understood compared with decarbonisation practices in many other sectors. A possible explanation is that agricultural GHG mitigation is primarily achieved through biophysical and biochemical processes, which might be much more complex and have larger uncertainties than abiotic sequestration and mitigation such as industrial carbon capture and storage (CCS) and other low-emission technical engineering changes (Fig. 10.4) (Smith et al. 2007; Smith and Olesen 2010; Moran and Wall 2011; Thamo et al. 2017, 2019; Lamb et al. 2022). That is probably why there are

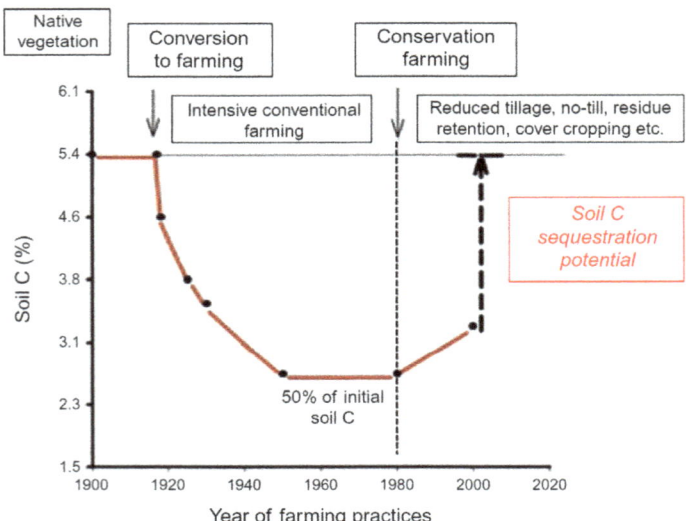

Fig. 10.3 Agricultural soil carbon sequestration. *Source* Chowdhury et al. (2021)

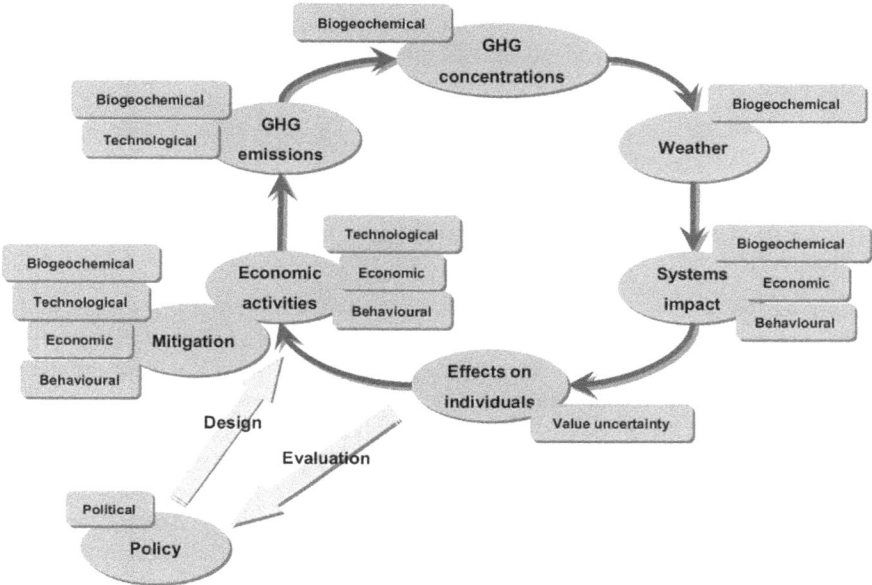

Fig. 10.4 Sources of uncertainty (in squares) in tackling climate change. *Source* Eory et al. (2019)

few technical choices for mitigating agricultural GHG emissions currently available. In addition, many important mechanisms behind agricultural carbon sequestration and GHG emissions reductions are not yet well known, e.g., the effects of agroforestry on N_2O and CH_4 emissions, the soil process behind N_2O emissions, and the underlying mechanisms for reducing GHG emissions from livestock urine and dung patches (Albrecht and Kandji 2003; Muller et al. 2011; Cai et al. 2017).

Both the magnitude of emissions and the relative lack of available mitigation technologies and knowledge make agricultural GHG emissions critical in tackling changing climate. To limit global warming to well below 2 °C and pursue efforts to limit the temperature increase to 1.5 °C above pre-industrial levels, it is necessary to achieve efficient carbon sequestration and GHG emissions reductions in agriculture. However, a range of challenges still exist. Agriculture is considerably less consolidated than other sectors; billions of people around the world need to take actions to tackle agricultural GHG emissions. Alongside climate goals, other targets must be met, right across the agricultural systems, from food security to biodiversity to the livelihood of rural communities (Smith et al. 2007; Smith and Olesen 2010; Tang et al. 2016b; Daigneault et al. 2018; Tang and Hailu 2020; Fujimori et al. 2022).

This chapter looks more deeply at the above issues by conducting a comprehensive review of the related literature. Specifically, this chapter examines strategies for agricultural carbon sequestration and GHG emissions reductions and their economic feasibility. The goal of this chapter is to provide a better understanding for policy-makers, professionals and researchers alike to stimulate the necessary changes in agriculture as a response to a changing climate.

10.2 Agricultural Strategies for Carbon Sequestration and GHG Emissions Reductions

Various agricultural strategies have been identified as appropriate measures to increase carbon sequestration and/or reduce GHG emissions (Fig. 10.5), offering opportunities for mitigating climate change. Generally, those strategies mitigate emissions through the mechanisms of reducing emissions, enhancing removals and avoiding emissions (Smith et al. 2008). It should be noted that the use of an agricultural strategy often impacts more than one gas by different mechanisms, implying that the net benefit evaluation is based on the integrated effects on various GHG gases (Smith and Olesen 2010; Moran and Wall 2011; Chowdhury et al. 2021). Moreover, the effectiveness of those sequestration and reduction strategies are often influenced by environmental conditions such as temperature, water, aeration, and biodiversity (Tang and Hailu 2020; Windisch et al. 2022).

To mitigate agricultural CO_2 emissions, increasing soil carbon sequestration is essential. Soil carbon sequestration is the net effect of biomass input into soils and

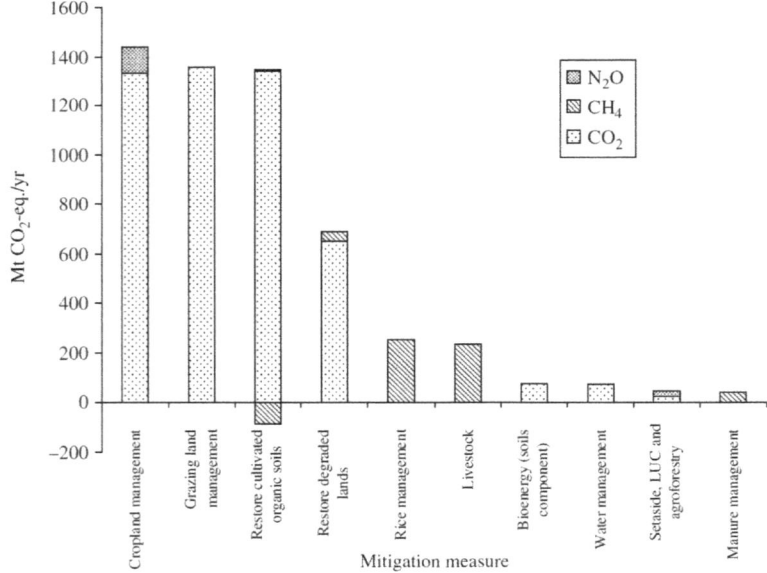

Fig. 10.5 Global biophysical mitigation potential for agricultural strategies by 2030. *Source* Smith and Olesen (2010)

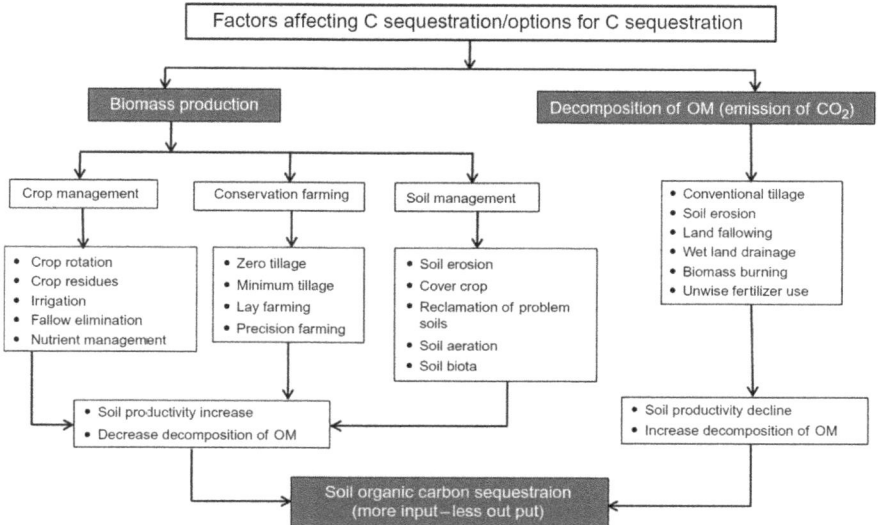

Fig. 10.6 Agricultural strategies affecting soil carbon sequestration. *Source* Chowdhury et al. (2021)

decomposition of soil organic matter; this means increasing soil carbon sequestration should be achieved through enhancing carbon input via biomass production and/or reducing carbon output via soil organic matter decomposition (Minasny et al. 2017; Chowdhury et al. 2021). In general, those two processes are influenced by agricultural practices substantially (Fig. 10.6). Environmental factors such as temperature and erosion and soil microorganisms condition also impact those processes (Six et al. 2006; Dębska et al. 2016; Chowdhury et al. 2021).

Conservation tillage, including no tillage and reduced tillage, has been considered to be an effective strategy to increase carbon stock in agricultural soils. Compared with conventional tillage, conservation tillage is much less likely to disturb soils and break down aggregates, thus playing a positive role in increasing soil carbon sequestration (Lal 2003; Baker et al. 2007; Minasny et al. 2017). Agricultural soils under conservative tillage are less vulnerable to erosion than soils under conventional tillage since the former have less-destroyed soil structure (Arshad et al. 1999; Cania et al. 2019). Additionally, global and regional studies have shown that the long-term use of conservation tillage improves biophysical and biochemical properties of agricultural soils, thereby inevitably enhancing soil quality and increasing organic carbon content in topsoils and/or plough layer (e.g., Liang et al. 2016; Chen et al. 2020; Mondal et al. 2020; Rahmati et al. 2020; Chowdhury et al. 2021; Krauss et al. 2022). Conservation tillage sustains or increases soil organic carbon when applied in intensive cropping enterprises (Reeves 1997). Besides soil organic carbon stock, other benefits brought by conservative tillage include improved organic matter quality and enhanced microbial diversity and activities (Chen et al. 2020; Li et al. 2020; Man et al. 2021).

Crop rotations influence the content and quality of agricultural soils and the yields of crops through the variations in soil carbon storage and aggregate stability and the increased biomass input, and therefore contribute to soil carbon sequestration (Persson et al. 2008; Jarecki et al. 2018). Moreover, using rotations with leguminous crops could reduce the nitrogen demand of subsequent crops due to the biological nitrogen fixation of leguminous crop, thereby reducing the GHG emissions of their production (Sanginga 2003; Tang et al. 2018, 2019). Empirical studies have shown that crop rotations with conservative tillage can have positive impacts on soil bacterial taxa, accelerate the accumulation rate of soil organic carbon and increase above- and below-ground residues for soil carbon input, thus increasing carbon sequestrated in soils and improving soil health (Reeves 1997; West and Post 2002; Yin et al. 2010; Deiss et al. 2021). Crop rotations with diverse crops might increase soil organic carbon by enhancing soil health condition (Russell et al. 2006; McDaniel et al. 2014) and enhance crop water use efficiency by increasing soil water content and reducing evaporation and surface water runoff (Pala et al. 2007), which would benefit to climate resilience increase and fragility decrease of agroecosystems (Yu et al. 2022).

Continuous cropping increases soil organic carbon via increasing residues inputs and reducing soil erosion, which can decrease soil organic matter decomposition and restrain soil organic carbon oxidation (Chowdhury et al. 2021). Moreover, continuous cropping may balance nutrient immobilisation and mineralisation, reducing nutrition loss in soils (Sherrod et al. 2003). In dryland systems, summer fallow with greater soil moisture and temperature conditions is likely to accelerate soil organic carbon oxidation and disrupt the balance between nutrient immobilisation and mineralisation, resulting in the decreasing levels of carbon and nitrogen in agricultural soils. Empirical studies have shown that the potential to sequester carbon with continuous cropping is larger than that with fallow system (Sherrod et al. 2003; Campbell et al. 2005; VandenBygaart et al. 2008). Even under conventional tillage, continuous cropping may still lead to an increase in soil carbon stocks (Machado et al. 2006). Continuous irrigated rice cropping with optimal fertiliser use can maintain or increase soil organic carbon and total nitrogen somewhat (Pampolino et al. 2008; Haefele et al. 2011).

The mean carbon content of dry crop residues is about 45% (Lal 1997). The carbon inputs for agricultural soils primarily come from the crop residues which remain in soils after harvest (Yang et al. 2013). Residue retention not only can maintain or improve soil organic carbon but also help to enhance soil aggregation, stabilise soil structure, decrease soil erosion, reduce evaporation, and increase soil nutrition condition (Duiker and Lal 1999; Whitbread et al. 2003; Kragt et al. 2012; Powlson et al. 2016; Kumar et al. 2019). Residue retention increases soil carbon concentration in topsoils (Lam et al. 2013; Sun et al. 2020), and the effect could be larger when residue retention is applied with appropriate strategies such as conservative tillage, crop rotations and improved fertiliser use (Yadav et al. 2019; Chowdhury et al. 2021). Turmel et al. (2015) have shown that in many developing countries residue retention with conservative tillage can additionally influence the effect of crop residue retention on organic carbon in the soil profile. In China, the large-scale implementation of crop straw/stover return incentivised by the authorities contributed largely to the soil organic carbon sequestration in the 2000s, with an average net increase of 0.14 tonne carbon per hectare per year (Zhao et al. 2018).

Improved fertilisation is believed to contribute to carbon sequestration and GHG emissions reductions. In low fertile soils, enhancing nutrition condition via improved fertilisation effectively increases biomass input and subsequently increases soil organic carbon content (Schipper et al. 2007). Applying both inorganic and organic fertilisers can promote biomass production through improving water use efficiency (Fan et al. 2005; Hati et al. 2006; Li et al. 2010), optimise carbon–nitrogen ratios of residues returned to the soils (Hijbeek et al. 2019), and reduce soil organic matter oxidation (Nayak et al. 2007; Srinivasarao et al. 2012), thereby effectively increasing soil carbon stocks and reducing GHG emissions (Vanlauwe et al. 2011; Ding et al. 2012; Das and Adhya 2014; Chang et al. 2020; Yuan et al. 2021, 2022). Biochar-based fertilisers are helpful in reducing leaching-induced nutrient loss and improving soil health condition, thus benefiting the environment through reducing GHG emissions (Niu et al. 2017; Grutzmacher et al. 2018; Osman et al. 2022). Improved fertilisation with farmyard manure could enhance soil nutrients and soil organic carbon in the Chinese Mollisols (Liu et al. 2006). Applying improved fertilisation which can improve nutrient use efficiency also tends to reduce nutrient loss and the associated N_2O emissions (Zhang et al. 2013; Powlson et al. 2018; Thilakarathna et al. 2020).

It has been widely believed that afforestation is an essential strategy to tackle global warming (e.g., Niu and Duiker 2006; Wolf et al. 2011; Nijnik and Halder 2013; Gao et al. 2014; Ovando et al. 2016; Yosef et al. 2018; Duffy et al. 2020; Eriksson 2020). Trees planted remove CO_2 from the atmosphere through photosynthesis and substantially increase carbon stocked in terrestrial ecosystems, thereby contributing to controlling atmospheric CO_2 concentrations. Plant cover changes brought by afforestation impact litter inputs and microbial community structure and activity in soils, enhancing the protection of soil organic matter against decomposition and oxidation (Jastrow et al. 2007; Garcia-Franco et al. 2015). Compared with other climate zones, carbon sequestration through afforestation in the tropics is quicker because of favourable growing factors such as temperature, water, growing season length, etc. (Arora and Montenegro 2011; Wolf et al. 2011; Ameray et al. 2021). Tropical forests have higher mean aboveground carbon density and larger carbon sequestration potential in comparison with subtropical, temperate and boreal ones (Bunker et al. 2005; Raihan et al. 2019; Ameray et al. 2021). In the subtropical region, using mixtures of native tree species is more likely to reduce nitrogen losses and have higher carbon stocks and fluxes than the use of monocultures (Lang et al. 2014; Liu et al. 2018). On China's semi-arid temperate Loess Plateau, afforestation on higher slope gradients and under shady slopes is likely to maintain more soil carbon (Zhang et al. 2020).

10.3 The Economic Feasibility of Agricultural Strategies for Carbon Sequestration and GHG Emissions Reductions[2]

Technically, the potential of carbon sequestration and GHG emissions reductions in agricultural systems is enormous (Lal 1999; Bernoux et al. 2006; Minasny et al. 2017; Duffy et al. 2020; Crippa et al. 2021). From an economic perspective, one expects farmers to only apply agricultural strategies for carbon sequestration and GHG emissions reductions if the benefits of the strategies are larger than their costs (Tang et al. 2016b, 2021). Therefore, it is necessary to comprehensively evaluate the economic impacts of those agricultural strategies. It should be noted that the economic impacts of those sequestration and reduction strategies are often influenced by multiple factors such as the studied region, the farming system type, biophysical approached adopted, economic approaches applied, and scenario assumptions.

Switching from convention to conservative tillage might be economically feasible for sequestrating carbon. In the Kansas River Valley in the central United States (US), no-tillage corn systems have larger soil carbon sequestration rate and higher net returns (1.51–2.87t CO_2e and $166–280 per hectare annually) than conventional-tillage corn systems in the absence of carbon credit markets (Pendell et al. 2007). In the corn-based systems in the central US, increasing the use of conservative tillage could sequestrate approximately 22.73 million tonnes CO_2e per year at a carbon price of $96.24/t CO_2e, and about 11.37 million tonnes CO_2e could be sequestrated annually at a carbon price of $24.06/t CO_2e (Antle et al. 2007). In Indiana US, farmers require an annual payment of $128.10 per hectare to switch from conventional tillage to no-till (Gramig and Widmar 2018). In Australia, a carbon price higher than $46.88/t CO_2e is likely to make conservative tillage economically viable (Lam et al. 2013). In the mainland southeastern Australia, the annual average net carbon sequestration rate for reduced tillage is 0.66t CO_2e per hectare, and that for no tillage is about 1.15t CO_2e per hectare. With a carbon price of $55.88/t CO_2e, about 1.56 million and 6.38 million tonnes CO_2e would be sequestrated over 20 years in this region with reduced tillage and no tillage, respectively (Grace et al. 2010). In the Indo-Gangetic Plain, no-till rice–wheat cropping systems could sequestrate 11 million tonnes CO_2e (7% of the sequestration potential) with a carbon price of $10.03/t CO_2e over 20 years, increasing to 26.77 million tonnes CO_2e (17% of the sequestration potential) with a carbon price of $20.05/t CO_2e (Grace et al. 2012). A carbon price of $10.03/t CO_2e could achieve 0.92 million tonnes CO_2e (14% of the sequestration potential) in no-till maize-wheat systems and 0.63 million tonnes CO_2e (6% of the sequestration potential) in no-till cotton-wheat systems over 20 years. A carbon price of $20.05/t CO_2e could realise 2.97 million tonnes CO_2e (35% of the sequestration potential) in

---

[2] All prices in Sect. 10.3 have been converted to 2022 US dollar as follows: First all prices are converted into US Dollar in the base year mentioned by the analysis by using the annual average closing exchange rate; then those dollar prices are converted to 2022 US dollars using the consumer price index from the US Bureau of Labor Statistics.

no-till maize-wheat systems and 1.89 million tonnes CO_2e (18% of the sequestration potential) in no-till cotton-wheat systems over 20 years (Grace et al. 2012).

The economic feasibility for crop rotations varies across regions. In the Midwest US, the sequestration cost for corn-soybean rotation could reach $567.27/t CO_2e with an annual sequestration rate of 0.075 tonne CO_2e per hectare (Choi and Sohngen 2010). In semi-arid western Senegal, the mitigation cost of a millet-groundnuts rotation is higher than $160.85/t CO_2e (Tschakert 2004). In the Central Wheatbelt of Western Australia, the mean cost of rotations comprising combinations of 10 crops is more than $64.35/t CO_2e (Kragt et al. 2012). In China's southern Jiangsu, adopting optimised rotation (clover-tomato-lettuce) in intensive vegetable cropping system could reduce N_2O emissions by 24% and increase the net profit by 33% at a carbon price of $29.46/t CO_2e (Min et al. 2016). In China's Loess Plateau Region, adopting optimised crop-pasture rotations in a semi-arid crop-livestock farming system may reduce on-farm GHG emission by 16.6 and 33% with marginal abatement costs not higher than $18.92/t CO_2e and $28.37/t CO_2e (Tang et al. 2019). When annual rainfall declines less than 10 per cent in this region, the $28.37/t CO_2e mitigation cost is accompanied by 13–17% GHG abatement with the use of optimised crop-pasture rotations (Tang and Hailu 2020).

The economic potentials of continuous cropping and residue retention are different. Continuous cropping is likely to be economically feasible strategies for GHG mitigation in agricultural system. In the northwestern US, the marginal cost for switching to continuous cropping is about $24.80/t CO_2e (Antle et al. 2001, 2002). For smallholder farming systems in northern Ghana, the net present value of farm profits would be increased by between 2 and 32% with the use of continuous cropping (González-Estrada et al. 2008). In China's Loess Plateau Region, adopting optimised crop-pasture rotations could limit the economic costs of agricultural carbon tax for smallholder farmers (Tang and Hailu 2020). Economic estimates about residue retention for mitigating GHG emissions tend to be mixed, which makes it difficult to draw firm conclusions. In China, the costs of residue retention range from $8.03/t CO_2e to $37.96/t CO_2e across regions (Chen et al. 2019). However, some argued that the costs of straw returning in the rice–wheat cropping systems in eastern China range from $66.38-$98.76/t CO_2e (Xia et al. 2014).

Existing literature about the economic feasibility of improved fertilisation provides mixed findings. In northern New South Wales of Australia, ensuring the economic feasibility for the irrigated cotton cropping system with optimised nitrogen fertilisation means the minimum required carbon price is $17.25/t CO_2e (Welsh et al. 2015). In southeastern Queensland of eastern Australia, the mitigation cost in irrigated cotton cropping system is considerably higher than $10.71/t CO_2e (Maraseni and Kodur 2019). In northeastern Kansas, beef cattle manure systems could be effective with sufficient incentives (net returns ranging from $72.50 to $185.80 per hectare) (Pendell et al. 2007). In western Canada, the economic feasibility of improved fertilisation for GHG mitigation in a wheat-pea cropping system is limited (Khakbazan et al. 2009).

In terms of afforestation, the developed and developing countries tend to have differing characteristics on economic feasibility. In general, the economic variable costs of afforestation in developing countries are relatively lower than the costs in developed countries, which might be related to low land and labour costs. For developing countries, the costs range from $2.53/t CO_2e in China (Hou et al. 2019) to $32.47/t CO_2e in Cameroon (Yaron, 2001), and many of them are less than $10/t CO_2e (Table 10.1). For developed countries, the costs are likely to be higher, ranging from $4.02 to $176.33/t CO_2e, and the majority of them are higher than $20/t CO_2e (Table 10.2). Some studies have reported that afforestation in tropical regions might be more cost-effective than in other regions (Raihan et al. 2019; Eriksson 2020). Though some studies argue that the economic feasibility of afforestation is limited, most researchers demonstrate that afforestation is more economically feasible compared with other mitigation strategies. Notwithstanding, the wide range of costs implies that it is uncertain at what scale and at what costs afforestation is feasible.

10.4   Discussion

While a growing body of literature has analysed the agricultural strategies for carbon sequestration and GHG emissions reductions and their economic feasibility, there are still issues that need more investigation.

While the co-benefits of the adoption of agricultural strategies for carbon sequestration and GHG emissions reductions have been documented in literature (e.g., Pala et al. 2007; Chen et al. 2020; Chowdhury et al. 2021; Man et al. 2021), the existing body of literature on those strategies' economic feasibility has rarely included those co-benefits in an integrated analytical framework. Studies tend to concentrate on carbon sequestration and/or GHG emissions reductions. In practice, the co-benefits of the adoption of agricultural mitigation strategies, such as improved soil health and productivity, improved water use efficiency, better management of erosion and salinity as well as positive biodiversity, are a non-negligible factor that drives farmers to mitigate and adapt to climate change (Dumbrell et al. 2016; Kragt et al. 2016; Fleming et al. 2019). To better promote the adoption of agricultural mitigation strategies, more considerations of co-benefits should be integrated into relevant climate policies. It is beneficial to recognise and value farmers for the multiple gains they generate through their adoption. A potential barrier is the difficulties in properly evaluating some co-benefits and the extent of co-benefits generated due to limited scientific and social knowledge, which should be well addressed by future research.

Agricultural climate policies are vulnerable to the complicated, short-term and cyclical nature of the politics which is subject to contradictions between diverse interests. In policy system, agricultural climate policies are easily outweighed by concerns about the livelihoods loss, food security, socioeconomic inequalities and other issues (Brown et al. 2019; Tang et al. 2019; Tang and Ma, 2022). In countries with relatively poorly functioning institution, such conflicts among policies might be especially hard to tackle (Henry and Tysiachniouk 2018; Turubanova et al. 2018). There is, however, a sparse body of literature on this issue. This shortcoming implies that the estimates of agricultural climate policies obtained in the literature may be effective under

Table 10.1 Summary of studies that have estimated carbon mitigation costs for afforestation in developing countries

Study	Region	Studied agroecosystem	Mitigation cost ($/t CO_2e)
De Jong et al. (2000)	The Central Highlands of Chiapas, southern Mexico	Subtropical to temperate, subhumid agriculture	25.63
Yaron (2001)	Mount Cameroon region, Cameroon	Tropic rainforest agriculture	32.47
Zelek and Shively (2003)	Mindanao Philippines	Tropical agriculture	15.61
Nijnik (2004)	Ukraine	Wastelands and low-profit agricultural lands	8.46
Olschewski et al. (2005)	Argentinean Patagonia	n. a	9.65
Wise and Cacho (2005)	Sumatra, Indonesia	Tropical tree–crop agroforest	15.48
Bellassen and Gitz (2008)	Congo Basin, Cameroon	Tropic rainforest agriculture	4.15
Fisher et al. (2011)	East and middle Tanzania	Savanna agriculture	8.48
Hoang et al. (2013)	Bac Kan Province, Vietnam	Subtropical agriculture	6.20
Yu et al. (2014)	Northern Shaanxi, China	Semi-arid agroforest (*R. pseudoacacia*)	9.84
Djanibekov and Khamzina (2016)	Khorezm region and southern districts of Karakalpakstan, Uzbekistan	Arid continental irrigated agriculture (*E. angustifolia*)	< 6.34
Hou et al. (2019)	China	Two-harvest rice with Eucalyptus	2.53
Kucuker (2019)	Korucu, Balıkesir Province, Turkey	Mediterranean mountainous agroforest	23.11
Weng et al. (2021)	China	n.a	20.86

Notes (1) Some studies only provide incentive prices rather than calling these direct mitigation costs. This chapter assumes that the lowest acceptable incentive price equals to the mitigation cost. (2) All costs have been converted to 2022 US dollar as follows: First all prices are converted into US Dollar in the base year mentioned by the analysis by using the annual average closing exchange rate; then those dollar prices are converted to 2022 US dollars using the consumer price index from the US Bureau of Labor Statistics. (3) n. a. = not available

Table 10.2 Summary of studies that have estimated carbon mitigation costs for afforestation in developed countries

Study	Region	Studied agroecosystem	Mitigation cost ($/t CO$_2$e)
Parks and Hardie (1995)	USA	n. a	51.51
Stavins (1999)	South central USA	n. a	62.75
Cacho et al. (2003)	Gippsland and Mount Gambier, Australia	Temperate tree–crop agroforest	31.86
Flugge and Abadi (2006)	Wheatbelt, Western Australia	Dryland grain production system	44.80
Hunt (2008)	North Queensland Australia	Wet tropic agriculture	62.67
Paul et al. (2013)	Southern Australia	Temperate agroforest	26.38
Polglase et al. (2013)	Australian cleared land area	Tropical, subtropical and temperate agroforest systems	52.76
Thamo et al. (2013)	Wheatbelt, Western Australia	Dryland grain production system	62.01
Nielsen et al. (2014)	USA	n. a	44.11
Anderson et al. (2015)	Alberta and British Columbia, Canada	n. a	176.33 (Alberta) 11.83(British Columbia)
Haim et al. (2016)	South Central and South East Regions, USA	n. a	38.88
Monge et al. (2016)	USA	n. a	13.32
Manley (2018, 2020)	New Zealand	Temperate agroforest (radiata pine)	8.32–10.81
Regan et al. (2020)	Southern South Australia	Broadacre agriculture (mallee)	37.36
West et al. (2020)	New Zealand	Temperate agroforest	10.27
Kovacs et al. (2021)	Lower Mississippi River Basin, Arkansas, USA	Humid continental climate agroforest	4.02

Notes (1) Some studies only provide incentive prices rather than calling these direct mitigation costs. This chapter assumes that the lowest acceptable incentive price equals to the mitigation cost. (2) All costs have been converted to 2022 US dollar as follows: First all prices are converted into US Dollar in the base year mentioned by the analysis by using the annual average closing exchange rate; then those dollar prices are converted to 2022 US dollars using the consumer price index from the US Bureau of Labor Statistics. (3) n. a. = not available

a much narrower range of conditions than expected by the policy designers. We need to further improve our recognition and understanding of the policy-making via developing better and integrated decision-making models (i.e., agent-based models) and comparing a wide range of relevant cases, thereby substantially increasing the effectiveness and robustness of agricultural climate policies.

10.5 Conclusions

This chapter conducts a comprehensive review of the literature related to the role of agriculture in achieving global net-zero emissions. Specifically, this chapter examines strategies for agricultural carbon sequestration and GHG emissions reductions and their economic feasibility. A general lesson from this review is that carbon sequestration and GHG emissions reductions in agriculture is potentially attractive, depending on the environmental conditions, socioeconomic contexts and strategies analysed.

Technically, the potential of carbon sequestration and GHG emissions reductions in agricultural systems is enormous. Various agricultural strategies have been identified as appropriate measures to increase carbon sequestration and/or reduce GHG emissions, offering opportunities for mitigating climate change. The effective strategies include conservation tillage, crop rotations, continuous cropping, residue retention, improved fertilisation, and afforestation.

Results on the economic feasibility of agricultural strategies for carbon sequestration and GHG emissions reductions are different. Adopting conservative tillage and continuous cropping might be economically feasible for sequestrating carbon. The economic feasibility for crop rotations varies across regions. Studies on the economic feasibility of improved fertilisation and residue retention provide mixed findings. The economic variable costs of afforestation in developing countries are relatively lower than the costs in developed countries.

While a growing body of related literature exists, there are still issues that need more investigation. More considerations of co-benefits should be integrated into relevant climate policies. It is beneficial to recognise and value farmers for the multiple gains they generate through their adoption. We need to further improve our recognition and understanding of the policy-making of agricultural climate policies, thereby substantially increasing their effectiveness and robustness.

References

Ahmed, J., Almeida, E., Aminetzah, D., Denis, N., Henderson, K., Katz, J., Kitchel, H., & Mannion, P. (2020). Agriculture and climate change: Reducing emissions through improved farming practices. McKinsey &Company. https://www.mckinsey.com/~/media/mckinsey/ind ustries/agriculture/our%20insights/reducing%20agriculture%20emissions%20through%20impr oved%20farming%20practices/agriculture-and-climate-change.pdf

Albrecht, A., & Kandji, S. T. (2003). Carbon sequestration in tropical agroforestry systems. *Agriculture, Ecosystems & Environment, 99*(1-3), 15-27.

Ameray, A., Bergeron, Y., Valeria, O., Montoro Girona, M., & Cavard, X. (2021). Forest carbon management: A review of silvicultural practices and management strategies across boreal, temperate and tropical forests. *Current Forestry Reports, 7*(4), 245-266.

Anderson, J. A., Long, A., & Luckert, M. K. (2015). A financial analysis of establishing poplar plantations for carbon offsets using Alberta and British Columbia's afforestation protocols. *Canadian Journal of Forest Research, 45*(2), 207-216.

Antle, J. M., Capalbo, S. M., Mooney, S., Elliott, E. T., & Paustian, K. H. (2001). Economic analysis of agricultural soil carbon sequestration: An integrated assessment approach. *Journal of Agricultural and Resource Economics, 26*(2): 344-367.

Antle, J., Capalbo, S., Mooney, S., Elliott, E. T., & Paustian, K. H. (2002). Sensitivity of carbon sequestration costs to soil carbon rates. *Environmental Pollution, 116*(3), 413-422.

Antle, J. M., Capalbo, S. M., Paustian, K., & Ali, M. K. (2007). Estimating the economic potential for agricultural soil carbon sequestration in the Central United States using an aggregate econometric-process simulation model. *Climatic Change, 80*(1), 145-171.

Arora, V. K., & Montenegro, A. (2011). Small temperature benefits provided by realistic afforestation efforts. *Nature Geoscience, 4*(8), 514-518.

Arshad, M. A., Franzluebbers, A. J., & Azooz, R. H. (1999). Components of surface soil structure under conventional and no-tillage in northwestern Canada. *Soil and Tillage Research, 53*(1), 41-47.

Baker, J. M., Ochsner, T. E., Venterea, R. T., & Griffis, T. J. (2007). Tillage and soil carbon sequestration: What do we really know?. *Agriculture, Ecosystems & Environment, 118*(1-4), 1-5.

Bellassen, V., & Gitz, V. (2008). Reducing emissions from deforestation and degradation in Cameroon—assessing costs and benefits. *Ecological Economics, 68*(1), 336-344.

Bennetzen, E. H., Smith, P., & Porter, J. R. (2016). Decoupling of greenhouse gas emissions from global agricultural production: 1970–2050. *Global Change Biology, 22*(2), 763-781.

Brown, C., Alexander, P., Arneth, A., Holman, I., & Rounsevell, M. (2019). Achievement of Paris climate goals unlikely due to time lags in the land system. *Nature Climate Change, 9*(3), 203-208.

Bunker, D. E., DeClerck, F., Bradford, J. C., Colwell, R. K., Perfecto, I., Phillips, O. L., Sankaran, M., & Naeem, S. (2005). Species loss and aboveground carbon storage in a tropical forest. *Science, 310*(5750), 1029-1031.

Cacho, O. J., Hean, R. L., & Wise, R. M. (2003). Carbon-accounting methods and reforestation incentives. *Australian Journal of Agricultural and Resource Economics, 47*(2), 153-179.

Cai, Y., Chang, S. X., & Cheng, Y. (2017). Greenhouse gas emissions from excreta patches of grazing animals and their mitigation strategies. *Earth-Science Reviews, 171*, 44-57.

Campbell, C. A., Janzen, H. H., Paustian, K., Gregorich, E. G., Sherrod, L., Liang, B. C., & Zentner, R. P. (2005). Carbon storage in soils of the North American Great Plains: Effect of cropping frequency. *Agronomy Journal, 97*(2), 349-363.

Cania, B., Vestergaard, G., Krauss, M., Fliessbach, A., Schloter, M., & Schulz, S. (2019). A long-term field experiment demonstrates the influence of tillage on the bacterial potential to produce soil structure-stabilizing agents such as exopolysaccharides and lipopolysaccharides. *Environmental Microbiome, 14*(1), 1-14.

Chang, N., Zhai, Z., Li, H., Wang, L., & Deng, J. (2020). Impacts of nitrogen management and organic matter application on nitrous oxide emissions and soil organic carbon from spring maize fields in the North China Plain. *Soil and Tillage Research*, 196, 104441.

Chen, H., Dai, Z., Veach, A. M., Zheng, J., Xu, J., & Schadt, C. W. (2020). Global meta-analyses show that conservation tillage practices promote soil fungal and bacterial biomass. *Agriculture, Ecosystems & Environment*, 293, 106841.

Chen, J., Gong, Y., Wang, S., Guan, B., Balkovic, J., & Kraxner, F. (2019). To burn or retain crop residues on croplands? An integrated analysis of crop residue management in China. *Science of The Total Environment*, 662, 141-150.

Choi, S. W., & Sohngen, B. (2010). The optimal choice of residue management, crop rotations, and cost of carbon sequestration: Empirical results in the Midwest US. *Climatic Change*, 99(1), 279-294.

Chowdhury, S., Bolan, N., Farrell, M., Sarkar, B., Sarker, J. R., Kirkham, M. B., Hossain, M.Z., & Kim, G. H. (2021). Role of cultural and nutrient management practices in carbon sequestration in agricultural soil. *Advances in Agronomy*, 166, 131-196.

Crippa, M., Solazzo, E., Guizzardi, D., Monforti-Ferrario, F., Tubiello, F. N., & Leip, A. J. N. F. (2021). Food systems are responsible for a third of global anthropogenic GHG emissions. *Nature Food*, 2(3), 198-209.

Daigneault, A., Greenhalgh, S., & Samarasinghe, O. (2018). Economic impacts of multiple agro-environmental policies on New Zealand land use. *Environmental and Resource Economics*, 69(4), 763-785.

Das, S., & Adhya, T. K. (2014). Effect of combine application of organic manure and inorganic fertilizer on methane and nitrous oxide emissions from a tropical flooded soil planted to rice. *Geoderma*, 213, 185-192.

Davis, S. J., Lewis, N. S., Shaner, M., et al. (2018). Net-zero emissions energy systems. *Science*, 360(6396), eaas9793.

De Jong, B. H., Tipper, R., & Montoya-Gómez, G. (2000). An economic analysis of the potential for carbon sequestration by forests: Evidence from southern Mexico. *Ecological Economics*, 33(2), 313-327.

Dębska, B., Długosz, J., Piotrowska-Długosz, A., & Banach-Szott, M. (2016). The impact of a bio-fertilizer on the soil organic matter status and carbon sequestration: Results from a field-scale study. *Journal of Soils and Sediments*, 16(10), 2335-2343.

Deiss, L., Sall, A., Demyan, M. S., & Culman, S. W. (2021). Does crop rotation affect soil organic matter stratification in tillage systems?. *Soil and Tillage Research*, 209, 104932.

Ding, X., Han, X., Liang, Y., Qiao, Y., Li, L., & Li, N. (2012). Changes in soil organic carbon pools after 10 years of continuous manuring combined with chemical fertilizer in a Mollisol in China. *Soil and Tillage Research*, 122, 36-41.

Djanibekov, U., & Khamzina, A. (2016). Stochastic economic assessment of afforestation on marginal land in irrigated farming system. *Environmental and Resource Economics*, 63(1), 95-117.

Duffy, C., O'Donoghue, C., Ryan, M., Styles, D., & Spillane, C. (2020). Afforestation: Replacing livestock emissions with carbon sequestration. *Journal of Environmental Management*, 264, 110523.

Duiker, S. W., & Lal, R. (1999). Crop residue and tillage effects on carbon sequestration in a Luvisol in central Ohio. *Soil and Tillage Research*, 52(1-2), 73-81.

Dumbrell, N. P., Kragt, M. E., & Gibson, F. L. (2016). What carbon farming activities are farmers likely to adopt? A best–worst scaling survey. *Land Use Policy*, 54, 29-37.

Eory, V., Topp, C. F., Butler, A., & Moran, D. (2018). Addressing uncertainty in efficient mitigation of agricultural greenhouse gas emissions. *Journal of Agricultural Economics*, 69(3), 627-645.

Eriksson, M. (2020). Afforestation and avoided deforestation in a multi-regional integrated assessment model. *Ecological Economics*, 169, 106452.

Fan, T., Stewart, B. A., Payne, W. A., Yong, W., Luo, J., & Gao, Y. (2005). Long-term fertilizer and water availability effects on cereal yield and soil chemical properties in northwest China. *Soil Science Society of America Journal*, 69(3), 842-855.

FAO (2022). FAOSTAT Emissions Totals, FAO. https://www.fao.org/faostat/en/#data/GT

Fisher, B., Lewis, S. L., Burgess, N. D., Malimbwi, R. E., Munishi, P. K., Swetnam, R. D., Kerry Turner, R., Willcock, S., & Balmford, A. (2011). Implementation and opportunity costs of reducing deforestation and forest degradation in Tanzania. *Nature Climate Change*, 1(3), 161-164.

Fleming, A., Stitzlein, C., Jakku, E., & Fielke, S. (2019). Missed opportunity? Framing actions around co-benefits for carbon mitigation in Australian agriculture. *Land Use Policy*, 85, 230-238.

Flugge, F., & Abadi, A. (2006). Farming carbon: an economic analysis of agroforestry for carbon sequestration and dryland salinity reduction in Western Australia. *Agroforestry Systems*, 68:181-192.

Fujimori, S., Wu, W., Doelman, J., Frank, S., Hristov, J., Kyle, P., Sands, R., Van Zeist, W.J., Havlik, P., Domínguez, I.P., & Takahashi, K. (2022). Land-based climate change mitigation measures can affect agricultural markets and food security. *Nature Food*, 3(2), 110-121.

Gao, Y., Zhu, X., Yu, G., He, N., Wang, Q., & Tian, J. (2014). Water use efficiency threshold for terrestrial ecosystem carbon sequestration in China under afforestation. *Agricultural and Forest Meteorology*, 195, 32-37.

Garcia-Franco, N., Martínez-Mena, M., Goberna, M., & Albaladejo, J. (2015). Changes in soil aggregation and microbial community structure control carbon sequestration after afforestation of semiarid shrublands. *Soil Biology and Biochemistry*, 87, 110-121.

González-Estrada, E., Rodriguez, L. C., Walen, V. K., Naab, J. B., Koo, J., Jones, J. W., Herrero, M., & Thornton, P. K. (2008). Carbon sequestration and farm income in West Africa: Identifying best management practices for smallholder agricultural systems in northern Ghana. *Ecological Economics*, 67(3), 492-502.

Grace, P. R., Antle, J., Aggarwal, P. K., Ogle, S., Paustian, K., & Basso, B. (2012). Soil carbon sequestration and associated economic costs for farming systems of the Indo-Gangetic Plain: A meta-analysis. *Agriculture, Ecosystems & Environment*, 146(1), 137-146.

Grace, P. R., Antle, J., Ogle, S., Paustian, K., & Basso, B. (2010). Soil carbon sequestration rates and associated economic costs for farming systems of south-eastern Australia. *Soil Research*, 48(8), 720-729.

Gramig, B. M., & Widmar, N. J. (2018). Farmer preferences for agricultural soil carbon sequestration schemes. *Applied Economic Perspectives and Policy*, 40(3), 502-521.

Grutzmacher, P., Puga, A. P., Bibar, M. P. S., Coscione, A. R., Packer, A. P., & de Andrade, C. A. (2018). Carbon stability and mitigation of fertilizer induced N_2O emissions in soil amended with biochar. *Science of the Total Environment*, 625, 1459-1466.

Gütschow, J., Jeffery L., & Gieseke, R. (2021). The PRIMAP-hist national historical emissions time series v2.3 (1850–2017), GFZ Data Services. https://doi.org/10.5880/pik.2019.001

Haefele, S. M., Konboon, Y., Wongboon, W., Amarante, S., Maarifat, A. A., Pfeiffer, E. M., & Knoblauch, C. J. F. C. R. (2011). Effects and fate of biochar from rice residues in rice-based systems. *Field Crops Research*, 121(3), 430-440.

Haim, D., White, E. M., & Alig, R. J. (2016). Agriculture afforestation for carbon sequestration under carbon markets in the United States: Leakage behavior from regional allowance programs. *Applied Economic Perspectives and Policy*, 38(1), 132-151.

Hati, K. M., Mandal, K. G., Misra, A. K., Ghosh, P. K., & Bandyopadhyay, K. K. (2006). Effect of inorganic fertilizer and farmyard manure on soil physical properties, root distribution, and water-use efficiency of soybean in Vertisols of central India. *Bioresource Technology*, 97(16), 2182-2188.

Henry, L. A., & Tysiachniouk, M. (2018). The uneven response to global environmental governance: Russia's contentious politics of forest certification. *Forest Policy and Economics*, 90, 97-105.

Hijbeek, R., Loon, M. P. V., & Ittersum, M. K. V. (2019). Fertiliser use and soil carbon sequestration: Trade-offs and opportunities. CCAFS Working Paper No 264. CGIAR Research Program on Climate Change (CCAFS).

Hoang, M. H., Do, T. H., Pham, M. T., van Noordwijk, M., & Minang, P. A. (2013). Benefit distribution across scales to reduce emissions from deforestation and forest degradation (REDD+) in Vietnam. *Land Use Policy*, 31, 48-60.

Hunt, C. (2008). Economy and ecology of emerging markets and credits for bio-sequestered carbon on private land in tropical Australia. *Ecological Economics*, 66(2), 309-318.

Jarecki, M., Grant, B., Smith, W., Deen, B., Drury, C., VanderZaag, A., Qian, B., Yang, J., & Wagner-Riddle, C. (2018). Long-term trends in corn yields and soil carbon under diversified crop rotations. *Journal of Environmental Quality*, 47(4), 635-643.

Jastrow, J. D., Amonette, J. E., & Bailey, V. L. (2007). Mechanisms controlling soil carbon turnover and their potential application for enhancing carbon sequestration. *Climatic Change*, 80(1), 5-23.

Khakbazan, M., Mohr, R. M., Derksen, D. A., et al. (2009). Effects of alternative management practices on the economics, energy and GHG emissions of a wheat–pea cropping system in the Canadian prairies. *Soil and Tillage Research*, 104(1), 30-38.

Kovacs, K. F., Haight, R. G., Moore, K., & Popp, M. (2021). Afforestation for carbon sequestration in the Lower Mississippi River Basin of Arkansas, USA: Does modeling of land use at fine spatial resolution reveal lower carbon cost?. *Forest Policy and Economics*, 130, 102526.

Köchy, M., Hiederer, R., & Freibauer, A. (2015). Global distribution of soil organic carbon–Part 1: Masses and frequency distributions of SOC stocks for the tropics, permafrost regions, wetlands, and the world. *Soil*, 1(1), 351-365.

Kragt, M. E., Gibson, F. L., Maseyk, F., & Wilson, K. A. (2016). Public willingness to pay for carbon farming and its co-benefits. *Ecological Economics*, 126, 125-131.

Kragt, M. E., Pannell, D. J., Robertson, M. J., & Thamo, T. (2012). Assessing costs of soil carbon sequestration by crop-livestock farmers in Western Australia. *Agricultural Systems*, 112, 27-37.

Krauss, M., Wiesmeier, M., Don, A., Cuperus, F., Gattinger, A., Gruber, S., Haagsma, W. K., Peigné, J., Palazzoli, M. C., Schulz, F., & Steffens, M. (2022). Reduced tillage in organic farming affects soil organic carbon stocks in temperate Europe. *Soil and Tillage Research*, 216, 105262.

Kucuker, D. M. (2019). Analyzing the effects of various forest management strategies and carbon prices on carbon dynamics in western Turkey. *Journal of Environmental Management*, 249, 109356.

Kumar, N., Nath, C. P., Hazra, K. K., Das, K., Venkatesh, M. S., Singh, M. K., Singh, S. S., Praharaj, C. S., & Singh, N. P. (2019). Impact of zero-till residue management and crop diversification with legumes on soil aggregation and carbon sequestration. *Soil and Tillage Research*, 189, 158-167.

Laborde, D., Mamun, A., Martin, W., Piñeiro, V., & Vos, R. (2021). Agricultural subsidies and global greenhouse gas emissions. *Nature Communications*, 12(1), 1-9.

Lal, R. (1997). Residue management, conservation tillage and soil restoration for mitigating greenhouse effect by CO_2-enrichment. *Soil and tillage research*, 43(1-2), 81-107.

Lal, R. (2003). Global potential of soil carbon sequestration to mitigate the greenhouse effect. *Critical Reviews in Plant Sciences*, 22(2), 151-184.

Lal, R., Negassa, W., & Lorenz, K. (2015). Carbon sequestration in soil. *Current Opinion in Environmental Sustainability*, 15, 79-86.

Lam, S. K., Chen, D., Mosier, A. R., & Roush, R. (2013). The potential for carbon sequestration in Australian agricultural soils is technically and economically limited. *Scientific Reports*, 3(1), 1-6.

Lamb, A., Green, R., Bateman, I., et al. (2016). The potential for land sparing to offset greenhouse gas emissions from agriculture. *Nature Climate Change*, 6(5), 488-492.

Lamb, W. F., Grubb, M., Diluiso, F., & Minx, J. C. (2022). Countries with sustained greenhouse gas emissions reductions: An analysis of trends and progress by sector. *Climate Policy*, 22(1), 1-17.

Lang, A. C., von Oheimb, G., Scherer-Lorenzen, M., Yang, B., Trogisch, S., Bruelheide, H., Ma, K., & Härdtle, W. (2014). Mixed afforestation of young subtropical trees promotes nitrogen acquisition and retention. *Journal of Applied Ecology*, 51(1), 224-233.

Li, F., Yu, J., Nong, M., Kang, S., & Zhang, J. (2010). Partial root-zone irrigation enhanced soil enzyme activities and water use of maize under different ratios of inorganic to organic nitrogen fertilizers. *Agricultural Water Management*, 97(2), 231-239.

Li, Y., Zhang, Q., Cai, Y., Yang, Q., & Chang, S. X. (2020). Minimum tillage and residue retention increase soil microbial population size and diversity: Implications for conservation tillage. *Science of the Total Environment*, 716, 137164.

Liang, A. Z., Yang, X. M., Zhang, X. P., Chen, X. W., Mclaughlin, N. B., Wei, S. C., Zhang, Y., Jia, S. X., & Zhang, S. X. (2016). Changes in soil organic carbon stocks under 10-year conservation tillage on a Black soil in Northeast China. *The Journal of Agricultural Science*, 154(8), 1425-1436.

Lin, Y., Ye, G., Kuzyakov, Y., Liu, D., Fan, J., & Ding, W. (2019). Long-term manure application increases soil organic matter and aggregation, and alters microbial community structure and keystone taxa. *Soil Biology and Biochemistry*, 134, 187-196.

Liu, X., Herbert, S. J., Hashemi, A. M., Zhang, X. F., & Ding, G. (2006). Effects of agricultural management on soil organic matter and carbon transformation-a review. *Plant Soil and Environment*, 52(12), 531.

Liu, X., Trogisch, S., He, J. S., Niklaus, P. A., Bruelheide, H., Tang, Z., Erfmeier, A., Scherer-Lorenzen, M., Pietsch, K. A., Yang, B., & Ma, K. (2018). Tree species richness increases ecosystem carbon storage in subtropical forests. *Proceedings of the Royal Society B*, 285(1885), 20181240.

Machado, S., Rhinhart, K., & Petrie, S. (2006). Long-term cropping system effects on carbon sequestration in eastern Oregon. *Journal of Environmental Quality*, 35(4), 1548-1553.

Man, M., Wagner-Riddle, C., Dunfield, K. E., Deen, B., & Simpson, M. J. (2021). Long-term crop rotation and different tillage practices alter soil organic matter composition and degradation. *Soil and Tillage Research*, 209, 104960.

Manley, B. (2018). Forecasting the effect of carbon price and log price on the afforestation rate in New Zealand. *Journal of Forest Economics*, 33, 112-120.

Manley, B. (2020). Impact on profitability, risk, optimum rotation age and afforestation of changing the New Zealand emissions trading scheme to an averaging approach. *Forest Policy and Economics*, 116, 102205.

Maraseni, T., & Kodur, S. (2019). Improved prediction of farm nitrous oxide emission through an understanding of the interaction among climate extremes, soil nitrogen dynamics and irrigation water. *Journal of Environmental Management*, 248, 109278.

McDaniel, M. D., Tiemann, L. K., & Grandy, A. S. (2014). Does agricultural crop diversity enhance soil microbial biomass and organic matter dynamics? A meta-analysis. *Ecological Applications*, 24(3), 560-570

Min, J., Lu, K., Sun, H., Xia, L., Zhang, H., & Shi, W. (2016). Global warming potential in an intensive vegetable cropping system as affected by crop rotation and nitrogen rate. *CLEAN–Soil, Air, Water*, 44(7), 766–774.

Minasny, B., Malone, B. P., McBratney, A. B., et al. (2017). Soil carbon 4 per mille. *Geoderma*, 292, 59-86.

Mondal, S., Chakraborty, D., Bandyopadhyay, K., Aggarwal, P., & Rana, D. S. (2020). A global analysis of the impact of zero-tillage on soil physical condition, organic carbon content, and plant root response. *Land Degradation & Development*, 31(5), 557-567.

Monge, J. J., Bryant, H. L., Gan, J., & Richardson, J. W. (2016). Land use and general equilibrium implications of a forest-based carbon sequestration policy in the United States. *Ecological Economics*, 127, 102-120.

Moran, D., & Wall, E. (2011). Livestock production and greenhouse gas emissions: Defining the problem and specifying solutions. *Animal Frontiers*, 1(1), 19-25.

Muller, A., Jawtusch, J., & Gattinger, A. (2011). Mitigating greenhouse gases in agriculture: A challenge and opportunity for agricultural policies. Diakonisches Werk der EKD e.V. for Brot für die Welt. https://orgprints.org/id/eprint/19989/1/gatti.pdf

Nayak, D. R., Babu, Y. J., Datta, A., & Adhya, T. K. (2007). Methane oxidation in an intensively cropped tropical rice field soil under long-term application of organic and mineral fertilizers. *Journal of Environmental Quality*, 36(6), 1577-1584.

Nielsen, A. S. E., Plantinga, A. J., & Alig, R. J. (2014). Mitigating climate change through afforestation: New cost estimates for the United States. *Resource and Energy Economics, 36*(1), 83-98.

Nijnik, M. (2004). Economics of climate change mitigation forest policy scenarios for Ukraine. *Climate Policy, 4*(3), 319-336.

Nijnik, M., & Halder, P. (2013). Afforestation and reforestation projects in South and South-East Asia under the Clean Development Mechanism: Trends and development opportunities. *Land Use Policy, 31*, 504-515.

Niu, X., & Duiker, S. W. (2006). Carbon sequestration potential by afforestation of marginal agricultural land in the Midwestern US. *Forest Ecology and Management, 223*(1-3), 415-427.

Niu, Y., Chen, Z., Müller, C., Zaman, M. M., Kim, D., Yu, H., & Ding, W. (2017). Yield-scaled N_2O emissions were effectively reduced by biochar amendment of sandy loam soil under maize-wheat rotation in the North China Plain. *Atmospheric Environment, 170*, 58-70.

Northrup, D. L., Basso, B., Wang, M. Q., Morgan, C. L., & Benfey, P. N. (2021). Novel technologies for emission reduction complement conservation agriculture to achieve negative emissions from row-crop production. *Proceedings of the National Academy of Sciences, 118*(28), e2022666118.

Olschewski, R., Benitez, P. C., De Koning, G. H. J., & Schlichter, T. (2005). How attractive are forest carbon sinks? Economic insights into supply and demand of certified emission reductions. *Journal of Forest Economics, 11*(2), 77-94.

Osman, A. I., Fawzy, S., Farghali, M., El-Azazy, M., Elgarahy, A. M., Fahim, R. A., Maksoud, M. I. A., Ajlan, A. A., Yousry, M., Saleem, Y., & Rooney, D. W. (2022). Biochar for agronomy, animal farming, anaerobic digestion, composting, water treatment, soil remediation, construction, energy storage, and carbon sequestration: A review. *Environmental Chemistry Letters.* https://doi.org/10.1007/s10311-022-01424-x

Ovando, P., Oviedo, J. L., & Campos, P. (2016). Measuring total social income of a stone pine afforestation in Huelva (Spain). *Land Use Policy, 50*, 479-489.

Pala, M., Ryan, J., Zhang, H., Singh, M., & Harris, H. C. (2007). Water-use efficiency of wheat-based rotation systems in a Mediterranean environment. *Agricultural Water Management, 93*(3), 136-144.

Pampolino, M. F., Laureles, E. V., Gines, H. C., & Buresh, R. J. (2008). Soil carbon and nitrogen changes in long-term continuous lowland rice cropping. *Soil Science Society of America Journal, 72*(3), 798-807.

Parks, P. J., & Hardie, I. W. (1995). Least-cost forest carbon reserves: Cost-effective subsidies to convert marginal agricultural land to forests. *Land Economics, 71*(1): 122-136.

Paul, K. I., Reeson, A., Polglase, P., Crossman, N., Freudenberger, D., & Hawkins, C. (2013). Economic and employment implications of a carbon market for integrated farm forestry and biodiverse environmental plantings. *Land Use Policy, 30*(1), 496-506.

Pendell, D. L., Williams, J. R., Boyles, S. B., Rice, C. W., & Nelson, R. G. (2007). Soil carbon sequestration strategies with alternative tillage and nitrogen sources under risk. *Applied Economic Perspectives and Policy, 29*(2), 247-268.

Persson, T., Bergkvist, G., & Kätterer, T. (2008). Long-term effects of crop rotations with and without perennial leys on soil carbon stocks and grain yields of winter wheat. *Nutrient Cycling in Agroecosystems, 81*(2), 193-202.

Polglase, P. J., Reeson, A., Hawkins, C. S., Paul, K. I., Siggins, A. W., Turner, J., Crawford, D. F., Jovanovic, T., Hobbs, T. J., Opie, K., & Almeida, A. (2013). Potential for forest carbon plantings to offset greenhouse emissions in Australia: economics and constraints to implementation. *Climatic Change, 121*(2), 161-175.

Powlson, D. S., Poulton, P. R., Macdonald, A. J., Johnston, A. E., White, R. P., & Goulding, K. W. T. (2018). 4 per mille—Is it feasible to sequester soil carbon at this rate annually in agricultural soils? In *Proceedings of the IFS Agronomic Conference*, Cambridge, UK, 6–7 December 2018.

Powlson, D. S., Stirling, C. M., Thierfelder, C., White, R. P., & Jat, M. L. (2016). Does conservation agriculture deliver climate change mitigation through soil carbon sequestration in tropical agro-ecosystems?. *Agriculture, Ecosystems & Environment, 220*, 164-174.

Rahmati, M., Eskandari, I., Kouselou, M., Feiziasl, V., Mahdavinia, G. R., Aliasgharzad, N., & McKenzie, B. M. (2020). Changes in soil organic carbon fractions and residence time five years after implementing conventional and conservation tillage practices. *Soil and Tillage Research*, 200, 104632.

Raihan, A., Begum, R. A., Mohd Said, M. N., & Abdullah, S. M. S. (2019). A review of emission reduction potential and cost savings through forest carbon sequestration. *Asian Journal of Water, Environment and Pollution*, 16(3), 1-7.

Reeves, D. W. (1997). The role of soil organic matter in maintaining soil quality in continuous cropping systems. *Soil and Tillage Research*, 43(1-2), 131-167.

Regan, C. M., Connor, J. D., Summers, D. M., Settre, C., O'Connor, P. J., & Cavagnaro, T. R. (2020). The influence of crediting and permanence periods on Australian forest-based carbon offset supply. *Land Use Policy*, 97, 104800.

Richards, M. B., Wollenberg, E., & Buglion-Gluck, S. (2015). *Agriculture's Contributions to National Emissions*. CGIAR Research Program on Climate Change, Agriculture and Food Security (CCAFS), Copenhagen.

Russell, A. E., Laird, D. A., & Mallarino, A. P. (2006). Nitrogen fertilization and cropping system impacts on soil quality in Midwestern Mollisols. *Soil Science Society of America Journal*, 70(1), 249-255.

Sanginga, N. (2003). Role of biological nitrogen fixation in legume based cropping systems; A case study of West Africa farming systems. *Plant and Soil*, 252(1), 25-39.

Sanz-Cobena, A., Lassaletta, L., Aguilera, E., del Prado, A., Garnier, J., Billen, G., Iglesias, A., Sanchez, B., Guardia, G., Abalos, D., & Smith, P. (2017). Strategies for greenhouse gas emissions mitigation in Mediterranean agriculture: A review. *Agriculture, Ecosystems & Environment*, 238, 5-24.

Saunois, M., Stavert, A. R., Poulter, B., et al. (2020). The global methane budget 2000–2017. *Earth System Science Data*, 12(3), 1561-1623.

Schipper, L. A., Baisden, W. T., Parfitt, R. L., Ross, C., Claydon, J. J., & Arnold, G. (2007). Large losses of soil C and N from soil profiles under pasture in New Zealand during the past 20 years. *Global Change Biology*, 13(6), 1138-1144.

Sherrod, L. A., Peterson, G. A., Westfall, D. G., & Ahuja, L. R. (2003). Cropping intensity enhances soil organic carbon and nitrogen in a no-till agroecosystem. *Soil Science Society of America Journal*, 67(5), 1533-1543.

Six, J., Frey, S. D., Thiet, R. K., & Batten, K. M. (2006). Bacterial and fungal contributions to carbon sequestration in agroecosystems. *Soil Science Society of America Journal*, 70(2), 555-569.

Smith, P., & Olesen, J. E. (2010). Synergies between the mitigation of, and adaptation to, climate change in agriculture. *The Journal of Agricultural Science*, 148(5), 543-552.

Smith, P., Martino, D., Cai, Z., Gwary, D., Janzen, H., Kumar, P., McCarl, B., Ogle, S., O'Mara, F., Rice, C., & Towprayoon, S. (2007). Policy and technological constraints to implementation of greenhouse gas mitigation options in agriculture. *Agriculture, Ecosystems & Environment*, 118(1-4), 6-28.

Smith, P., Martino, D., Cai, Z., Gwary, D., Janzen, H., Kumar, P., McCarl, B., Ogle, S., O'Mara, F., Rice, C., Scholes, B., Sirotenko, O., Howden, M., McAllister, T., Pan, G., Romanenkov, V., Schneider, U., Towprayoon, S., Wattenbach, M., & Smith, J. (2008). Greenhouse gas mitigation in agriculture. *Philosophical transactions of the Royal Society of London. Series B, Biological sciences*, 363(1492), 789–813.

Srinivasarao, C., Deshpande, A. N., Venkateswarlu, B., Lal, R., Singh, A. K., Kundu, S., Vittal, K. P. R., Mishra, P. K., Prasad, J. V. N. S., Mandal, U. K., & Sharma, K. L. (2012). Grain yield and carbon sequestration potential of post monsoon sorghum cultivation in Vertisols in the semi arid tropics of central India. *Geoderma*, 175, 90-97.

Stavins, R. N. (1999). The cost of carbon sequestration: A revealed-preference approach. *American Economic Review*, 89(4): 994-1009.

Sun, W., Canadell, J. G., Yu, L., Yu, L., Zhang, W., Smith, P., Fischer, T., & Huang, Y. (2020). Climate drives global soil carbon sequestration and crop yield changes under conservation agriculture. *Global Change Biology*, 26(6), 3325-3335.

Tang, K., & Hailu, A. (2020). Smallholder farms' adaptation to the impacts of climate change: Evidence from China's Loess Plateau. *Land Use Policy*, 91, 104353.

Tang, K., & Ma, C. (2022). The cost-effectiveness of agricultural greenhouse gas reduction under diverse carbon policies in China. *China Agricultural Economic Review*. https://doi.org/10.1108/CAER-01-2022-0008

Tang, K., Hailu, A., Kragt, M. E., & Ma, C. (2016a). Marginal abatement costs of greenhouse gas emissions: Broadacre farming in the Great Southern Region of Western Australia. *Australian Journal of Agricultural and Resource Economics*, 60(3), 459-475.

Tang, K., Kragt, M. E., Hailu, A., & Ma, C. (2016b). Carbon farming economics: What have we learned?. *Journal of Environmental Management*, 172, 49-57.

Tang, K., Hailu, A., Kragt, M. E., & Ma, C. (2018). The response of broadacre mixed crop-livestock farmers to agricultural greenhouse gas abatement incentives. *Agricultural Systems*, 160, 11-20.

Tang, K., He, C., Ma, C., & Wang, D. (2019). Does carbon farming provide a cost-effective option to mitigate GHG emissions? Evidence from China. *Australian Journal of Agricultural and Resource Economics*, 63(3), 575-592.

Tang, K., Wang, M., & Zhou, D. (2021). Abatement potential and cost of agricultural greenhouse gases in Australian dryland farming system. *Environmental Science and Pollution Research*, 28(17), 21862-21873.

Thamo, T., Addai, D., Kragt, M. E., Kingwell, R. S., Pannell, D. J., & Robertson, M. J. (2019). Climate change reduces the mitigation obtainable from sequestration in an Australian farming system. *Australian Journal of Agricultural and Resource Economics*, 63(4), 841-865.

Thamo, T., Addai, D., Pannell, D. J., Robertson, M. J., Thomas, D. T., & Young, J. M. (2017). Climate change impacts and farm-level adaptation: Economic analysis of a mixed cropping-livestock system. *Agricultural Systems*, 150, 99-108.

Thamo, T., Kingwell, R. S., & Pannell, D. J. (2013). Measurement of greenhouse gas emissions from agriculture: Economic implications for policy and agricultural producers. *Australian Journal of Agricultural and Resource Economics*, 57(2), 234-252.

Thilakarathna, S. K., Hernandez-Ramirez, G., Puurveen, D., Kryzanowski, L., Lohstraeter, G., Powers, L. A., Quan, N., & Tenuta, M. (2020). Nitrous oxide emissions and nitrogen use efficiency in wheat: Nitrogen fertilization timing and formulation, soil nitrogen, and weather effects. *Soil Science Society of America Journal*, 84(6), 1910-1927.

Tongwane, M. I., & Moeletsi, M. E. (2018). A review of greenhouse gas emissions from the agriculture sector in Africa. *Agricultural Systems*, 166, 124-134.

Tschakert, P. (2004). The costs of soil carbon sequestration: An economic analysis for small-scale farming systems in Senegal. *Agricultural Systems*, 81(3), 227-253.

Tubiello, F. N., Karl, K., Flammini, A., Gütschow, J., Conchedda, G., Pan, X., Qi, S.Y., Halldórudóttir Heiðarsdóttir, H., Wanner, N., Quadrelli, R., & Torero, M. (2022). Pre-and post-production processes increasingly dominate greenhouse gas emissions from agri-food systems. *Earth System Science Data*, 14(4), 1795-1809.

Turmel, M. S., Speratti, A., Baudron, F., Verhulst, N., & Govaerts, B. (2015). Crop residue management and soil health: A systems analysis. *Agricultural Systems*, 134, 6-16.

Turubanova, S., Potapov, P. V., Tyukavina, A., & Hansen, M. C. (2018). Ongoing primary forest loss in Brazil, Democratic Republic of the Congo, and Indonesia. *Environmental Research Letters*, 13(7), 074028.

VandenBygaart, A. J., McConkey, B. G., Angers, D. A., Smith, W., De Gooijer, H., Bentham, M., & Martin, T. (2008). Soil carbon change factors for the Canadian agriculture national greenhouse gas inventory. *Canadian Journal of Soil Science*, 88(5), 671-680.

Vanlauwe, B., Kihara, J., Chivenge, P., Pypers, P., Coe, R., & Six, J. (2011). Agronomic use efficiency of N fertilizer in maize-based systems in sub-Saharan Africa within the context of integrated soil fertility management. *Plant and Soil*, 339(1), 35-50.

Welsh, J., Powell, J., & Scott, F. (2015). Optimising nitrogen fertiliser in high yielding irrigated cotton: A benefit-cost analysis and the feasibility of participation in the ERF. *Australian Farm Business Management Journal*, 12, 62-80.

Weng, Y., Cai, W., & Wang, C. (2021). Evaluating the use of BECCS and afforestation under China's carbon-neutral target for 2060. *Applied Energy*, 299, 117263.

West, T. O., & Post, W. M. (2002). Soil organic carbon sequestration rates by tillage and crop rotation: A global data analysis. *Soil Science Society of America Journal*, 66(6), 1930-1946.

Whitbread, A., Blair, G., Konboon, Y., Lefroy, R., & Naklang, K. (2003). Managing crop residues, fertilizers and leaf litters to improve soil C, nutrient balances, and the grain yield of rice and wheat cropping systems in Thailand and Australia. *Agriculture, Ecosystems & Environment*, 100(2-3), 251-263.

Windisch, M. G., Humpenöder, F., Lejeune, Q., Schleussner, C. F., Lotze-Campen, H., & Popp, A. (2022). Accounting for local temperature effect substantially alters afforestation patterns. *Environmental Research Letters*, 17(2), 024030.

Wise, R., & Cacho, O. (2005). Tree-crop interactions and their environmental and economic implications in the presence of carbon-sequestration payments. *Environmental Modelling & Software*, 20(9), 1139-1148.

Wolf, S., Eugster, W., Potvin, C., Turner, B. L., & Buchmann, N. (2011). Carbon sequestration potential of tropical pasture compared with afforestation in Panama. *Global Change Biology*, 17(9), 2763-2780.

Wollenberg, E., Richards, M., Smith, P., Havlík, P., Obersteiner, M., Tubiello, F. N., Herold, M., Gerber, P., Carter, S., Reisinger, A., & Campbell, B. M. (2016). Reducing emissions from agriculture to meet the 2 °C target. *Global Change Biology*, 22(12), 3859-3864.

Xia, L., Wang, S., & Yan, X. (2014). Effects of long-term straw incorporation on the net global warming potential and the net economic benefit in a rice-wheat cropping system in China. *Agriculture, Ecosystems & Environment*, 197, 118-127.

Yadav, G. S., Lal, R., Meena, R. S., Babu, S., Das, A., Bhowmik, S. N., Datta, M., Layak, J., & Saha, P. (2019). Conservation tillage and nutrient management effects on productivity and soil carbon sequestration under double cropping of rice in north eastern region of India. *Ecological Indicators*, 105, 303-315.

Yang, X., Drury, C. F., & Wander, M. M. (2013). A wide view of no-tillage practices and soil organic carbon sequestration. *Acta Agriculturae Scandinavica, Section B-Soil & Plant Science*, 63(6), 523-530.

Yaron, G. (2001). Forest, plantation crops or small-scale agriculture? An economic analysis of alternative land use options in the Mount Cameroon area. *Journal of Environmental Planning and Management*, 44(1), 85-108.

Yin, C., Jones, K. L., Peterson, D. E., Garrett, K. A., Hulbert, S. H., & Paulitz, T. C. (2010). Members of soil bacterial communities sensitive to tillage and crop rotation. *Soil Biology and Biochemistry*, 42(12), 2111-2118.

Yosef, G., Walko, R., Avisar, R., Tatarinov, F., Rotenberg, E., & Yakir, D. (2018). Large-scale semi-arid afforestation can enhance precipitation and carbon sequestration potential. *Scientific Reports*, 8(1), 1-10.

Yu, J., Yao, S., & Zhang, B. (2014). Designing afforestation subsidies that account for the benefits of carbon sequestration: A case study using data from China's Loess Plateau. *Journal of Forest Economics*, 20(1), 65-76.

Yu, T., Mahe, L., Li, Y., Wei, X., Deng, X., & Zhang, D. (2022). Benefits of crop rotation on climate resilience and its prospects in China. *Agronomy*, 12(2), 436.

Yuan, F., Tang, K., & Shi, Q. (2021). Does Internet use reduce chemical fertilizer use? Evidence from rural households in China. *Environmental Science and Pollution Research*, 28(5), 6005-6017.

Yuan, F., Tang, K., Shi, Q., Qiu, W., & Wang, M. (2022). Rural women and chemical fertiliser use in rural China. *Journal of Cleaner Production*, 344, 130959.

Zelek, C. A., & Shively, G. E. (2003). Measuring the opportunity cost of carbon sequestration in tropical agriculture. *Land Economics*, 79(3), 342-354.

Zhang, W. F., Dou, Z. X., He, P., Ju, X. T., Powlson, D., Chadwick, D., Norse, D., Lu, Y. L., Zhang, Y., Wu, L., & Zhang, F. S. (2013). New technologies reduce greenhouse gas emissions from nitrogenous fertilizer in China. *Proceedings of the National Academy of Sciences*, 110(21), 8375-8380.

Zhang, X., Adamowski, J. F., Liu, C., Zhou, J., Zhu, G., Dong, X., Cao, J., & Feng, Q. (2020). Which slope aspect and gradient provides the best afforestation-driven soil carbon sequestration on the China's Loess Plateau?. *Ecological Engineering*, 147, 105782.

Zhao, Y., Wang, M., Hu, S., Zhang, X., Ouyang, Z., Zhang, G., Huang, B., Zhao, S., Wu, J., Xie, D., & Shi, X. (2018). Economics-and policy-driven organic carbon input enhancement dominates soil organic carbon accumulation in Chinese croplands. *Proceedings of the National Academy of Sciences*, 115(16), 4045-4050.

Milton Keynes UK
Ingram Content Group UK Ltd.
UKHW020610231023
431161UK00002B/4

9 789811 955648